PLANTS IN HAWAIIAN CULTURE

PLANTS
IN HAWAIIAN CULTURE

BEATRICE H. KRAUSS

ILLUSTRATIONS BY THELMA F. GREIG

For Lila
with much Aloha
and happy memories
of our frienship
Bea

aloha i ka aina
Thelma

Dec. 4, 1993

A KOLOWALU BOOK

UNIVERSITY OF HAWAII PRESS / HONOLULU

Printed in the United States of America
93 94 95 96 97 98 5 4 3 2 1

Library of Congress Cataloging-in-Publication Data
Krauss, Beatrice H.
Plants in Hawaiian culture / Beatrice H. Krauss ; illustrations by
Thelma F. Greig
p. cm.
"A Kolowalu book."
Includes bibliographical references and index.
ISBN 0-8248-1225-5 (alk. paper)
1. Hawaiians—Ethnobotany. 2. Ethnobotany—Hawaii.
3. Hawaiians—Material culture. 4. Plants, Useful—Hawaii.
I. Title.
DU624.65.K72 1993
581.6'1'09969—dc20 93-14789
CIP

The photographs were taken by Thelma Grieg, with the following exceptions: Frank Mitchell, Plates 4, 5, 25 (inset), 36, 37, 47, 65, and 66; Al Miller, Plates 8, 42, and 73 (inset), and David Boyenton, Plate 57.

Front cover: Tommy Kaulukukui with poi pounder and pounding board. (Photo by Thelma Greig.)

University of Hawaii Press books are printed on acid-free paper and meet the guidelines for permanence and durability of the Council on Library Resources

Designed by Kenneth Miyamoto

CONTENTS

PREFACE

THIS BOOK is concerned with ethnobotany, a comparatively new science. It involves the disciplines of ethnology, the study of native (that is, original) people, and of botany, the investigation of plants. Thus, ethnobotany constitutes the study of the interrelation of native people and plants or, stated in another way, the dependence of native people on plants, both wild and cultivated. In the case of Hawaiians, the dependence was paramount because they had neither metal for tools, nails, and utensils, nor did they know of ceramic clays for pottery.

The term *ethnobotany* implies a way of life without foreign influence. This book is therefore concerned only with the culture of the Hawaiians before the first known contact with foreigners, the arrival of Captain James Cook in 1778. This period is referred to as precontact or pre-European.

This book is intended as a general introduction to the ethnobotany of the Hawaiians and as such it presumes, on the part of the reader, little background in either botany or Hawaiian ethnology. It describes the plants themselves, whether cultivated or brought from the forests, streams, or ocean, as well as the modes of cultivation and collection. It discusses the preparation and uses of the plant materials, and the methods employed in building houses and making canoes, wearing apparel, and the many other artifacts that were part of the material culture associated with this farming and fishing people.

Because the Hawaiians had no written language, information on their culture was transmitted orally and has come to us through spoken legends, chants, and stories, passed from generation to generation and eventually recorded in postcontact times. Accounts by early foreign voyagers, observations by early settlers, including missionaries of many faiths, and studies by early authors such as Malo and Kamakau and such scholars as Buck, Handy, and Pukui have added to our knowledge. However, the reader is cautioned on the reliability of some observations and conclusions drawn by early visitors. A list of shortened references used in preparing this book appears at the end of each chapter; when no pagination is given in these references, the reference is an entire book or article. The reader interested in gaining a more detailed understanding of ancient Hawaiian culture should consult these references. A bibliography is included at the end of the book.

Every decision made, every task performed, every event, small or large, was accompanied by rituals and prayers, but many of these have been lost. Those that have survived have been drawn from the memories of older Hawaiians and, increasingly these days, are being passed on to younger Hawaiians who are becoming interested in this aspect of their heritage.

Underlying all ritual activity was *mana,* the fundamental principle that shaped and guided ancient Hawaiian spirituality. *Mana* was an integral part of the life of these ancient people. For them, *mana* implied a pervasive supernatural power as an essential part of their existence and that of the world around them. *Mana* involves mystic and psychic faculties that have been largely lost, although some

Hawaiians today are conscious of these forces and some younger Hawaiians are being made aware of them.

The role of *mana* and the rituals and prayers that attended so many phases of everyday life in precontact Hawai'i can hardly be overemphasized. Nevertheless, an introductory work such as this can hope to do little more than hint at this aspect of Hawaiian culture. It is hoped that this book, focused on the ethnobotanical aspects of the culture of old Hawai'i, will be followed by other books in which the spiritual dimensions are more fully conveyed.

Hawaiian terms with their English equivalents are given when a term is first introduced in the book; it is important that we all become familiar with these Hawaiian terms. Thereafter, Hawaiian and English words are used interchangeably.

Descriptions and photographs of most of the plants referred to in the book appear in a section at the end of the book. These photographs consist of habitat views of native plants and details of parts of these plants referred to in the text. Line drawings illustrate artifacts and details of canoes and houses. A glossary includes all Hawaiian, common, and scientific names of plants mentioned.

The source for the scientific names of plants is the *Manual of the flowering plants of Hawai'i,* vols. 1 and 2, by Warren L. Wagner, Derral R. Herbst, and S. H. Sohmer, published by University of Hawaii Press and Bishop Museum Press in 1990. For those plants not included in the *Manual,* the *Hawaiian dictionary,* by Mary Kawena Pukui and Samuel H. Elbert, published by University of Hawaii Press in 1986, was used. The sources for Hawaiian words are the *Hawaiian dictionary,* the original references for words that did not appear in that work, or Elbert (personal communication).

The term *native plant* is used in an ethnobotanical sense to include endemic (found only in Hawai'i), indigenous (found in Hawai'i and elsewhere, but unique in each habitat), and Polynesian-introduced plants. In strictly botanical circles, the last-named group is not considered native.

Descriptions of plants were obtained from the following sources: *Hawaii's vanishing flora* by Bert Y. Kimura and Kenneth M. Nagata, *Trailside plants of Hawaii's National Parks* by Charles H. Lamoureux, *In gardens of Hawaii* by Marie C. Neal, *Flora Hawaiiensis* by Otto (and Isa) Degener, *The indigenous trees of the Hawaiian Islands* by J. F. Rock, and *Manual of the flowering plants of Hawai'i,* cited above.

ACKNOWLEDGMENTS

I WAS MOST fortunate to have my longtime friend Thelma Greig take most of the photographs and make all of the line-drawing illustrations. The former were taken in the field and most of the latter drawn from artifacts in the Bishop Museum. My grateful thanks to Tony Han and her assistants at Bishop Museum for making those artifacts accessible. Thanks also to the Harold L. Lyon Arboretum, the Hawaii Maritime Center, and Kiope Mossman for allowing artifacts from their collections to be drawn.

To my editors Iris Wiley and Pat Crosby go my special thanks for their encouraging, supportive, long-suffering but patient nurturing that finally brought this project to completion. Thanks also to the copy editor Eileen D'Araujo, the associate managing editor, Cheri Dunn, and all those in the production department of the University of Hawaii Press who are responsible for the finished product.

I wish especially to express my gratitude to Derral Herbst for the many hours he spent examining the manuscript for accuracy in the identification of plants and their scientific nomenclature. I alone, of course, bear responsibility for any of this work's errors.

An early draft of relevant chapters of the text was submitted to each of the following experts: Russ Apple, houses; Tommy Holmes, canoes; Adrienne Kaeppler, clothing; the late Donald Mitchell, sports and games; and Barbara Smith, music and musical instruments. I am very grateful to each of these persons for critically reading these chapters, and for their suggestions.

Many thanks to Joyce Jacobson for guiding Thelma Greig and me to plants that were studied and photographed on the island of Hawai'i. I wish to acknowledge help from, among others, Robert Hirano of the Lyon Arboretum and staff members at the Waimea Arboretum and Botanical Garden, Al Miller and Joyce Tomlinson, for help in obtaining specimens of some of the plants.

Not forgotten is Toki Murakami, secretary at the Lyon Arboretum, who, with the aid of a series of student helpers, typed several of the five drafts of the manuscript. And finally I am grateful to all those people who, over many years, offered me encouragement and support.

INTRODUCTION

We now know that the Polynesians who settled in Hawai'i came from the Marquesas Islands in the fifth or sixth centuries or even earlier, but we are not sure where the first landing or settlement was made, nor do we know how they came to settle in Hawai'i. The most compelling reason probably is that the Polynesians' love of the sea, and their adventurous nature, spurred them on to find what lay beyond the horizon. Other incentives may have been overpopulation or warfare.

Whatever the incentive or incentives, the Polynesians, superb navigators, became aware that they were approaching a new land, recognizing well-known signs: changes in the ocean currents, the sight of land and sea birds, the green cast on the undersides of clouds. Then, still a hundred miles out at sea, they would have been able to see the peaks of Mauna Loa and Mauna Kea.

When these first voyagers landed and settled in Hawai'i, they would have found many things to their liking: no inhabitants to conquer or make peace with, ample fresh water in streams and springs, forested mountains, and an abundance of fish in both streams and ocean. On the other hand, there were no plants to use for food, although later they were to find a few, of little value, in the forests. However, an exploratory/settler group would have carried planting material for both food and other plants with them besides what they needed for consumption on the journey. There is some indication that there were return trips to their former homeland; on their return to Hawai'i they would have perhaps brought additional planting material.

These settlers first established settlements along the shoreline, preferably at the entrance to valleys. Here the mouth of the stream furnished fresh water, the ocean a plentiful supply of fish. As a stream nears its mouth in wide valleys, it becomes more shallow and spreading; this often makes for a marshy area ideal for growing taro, the settlers' main food crop. On nearby hillocks they planted sweet potatoes, and coconuts along the strand. Sugarcane was planted near houses, yams in nearby woods, breadfruit at the edge of these woods, and bananas in sheltered areas nearby.

Exploration of the forests revealed trees, the timber of which was valuable for building houses and making canoes. The forests also yielded plants that could be used for making and dying tapa, for medicine, and for a variety of other artifacts.

As the established settlements grew, and some additional migrations increased the population, it became necessary to move inland and up the valleys. Along the streams, small areas were cleared to plant taro, and hillsides nearby were prepared for planting other crops. Gradually the land along the length and across the width of the valley floors was cleared and planted to wet taro, irrigated with an elaborate system of ditches. The slopes lying below the high forests were, in time, also cleared and planted to crops that did not need irrigation, such as dry taro. Now there were dispersed communities in the valleys and on their slopes in addition to the original fishing settlements by the sea.

The Polynesians brought with them economic, social, and political practices from their homeland. These became modified and changed as a result of time, isolation, and local conditions. The social structure of the Hawaiians came to consist of nobility *(aliʻi),* the king, *mōʻī,* and his family, as well as chiefs; the king's advisors, "agents" or "cabinet," *ʻaha kuhina;* professionals and priests, both groups called *kāhuna,* with the priests having considerable power; and the great mass of commoners, *makaʻāinana,* who were primarily farmers. Slaves *(kauā)* were the lowest members of society.

The power of the king was great; he held power of life and death over his people and governed the use of land. However, he did not own land himself. Land belonged to the gods and he served as custodian of it: to see that it was cared for and cultivated and that the harvest was properly divided. He conferred with his cabinet in matters of governance, and had agents such as ones who collected taxes or procured fighting men in case of war. The system of governance could be characterized as feudal, resembling that found in central Europe in the Middle Ages. However, the European serf was bound to his "master" for life, but the Hawaiian *makaʻāinana* could move from master to master at will.

An island was divided into land divisions called *ahupuaʻa.* Such a division frequently consisted of a valley with the sea forming the *makai* boundary and the ridge of the mountain range at the head of the valley being the *mauka* limit. The crests of the ridges forming the sides of the valley completed the boundaries of that particular *ahupuaʻa.* The king appointed a chief to govern an *ahupuaʻa,* who in turn named *konohiki,* headmen, who were in charge of divisions within the *ahupuaʻa.* Under the headmen were the *luna* in charge of the farmers, the *makaʻāinana.* The story of this book is primarily that of these farmers and their fellow *makaʻāinana* who fished.

SOURCES

Buck (1965, 23–34); Emory (1965e, 319–322); Handy and Handy (1972, 8, 13, 18–20); Handy and Pukui (1972, 1–5); Handy and others (1965, 35–46); Kirch (1985, 1, 8, 58–60, 65–67, 87–88, 247–248, 284, 286–287); Wise (1965a).

1
FOOD

THE HAWAIIAN farmer, *mahi'ai,* was a horticulturist par excellence; would that the agriculturist in Hawai'i today were as skillful a planter! In this book it is possible to touch only briefly on the extent and character of the practical knowledge that the Hawaiians acquired, adopted, and utilized in their farming.

They demonstrated great ability in systematic differentiation, identification, and naming of the plants they cultivated and gathered for use. Their knowledge of the gross morphology of plants, their habits of growth, and the requirements for greatest yields is not excelled by expert agriculturists of more complicated cultures. They worked out the procedures of cultivation for every locality, for all altitudes, for different weather conditions and exposures, and for soils of all types. In their close observations of the plants they grew, they noted and selected mutants (sports) and natural hybrids, and so created varieties of the plants they already had. Thus over the years after their arrival in the Islands, the Hawaiians added hundreds of named varieties of taro, sweet potatoes, sugarcane, and other cultivated plants to those they had brought with them from the central Pacific.

They recognized the value of what today is known as "organic farming": incorporating into their cultivated areas such plant materials as weeds that grew during fallow periods and unused leafy portions of food plants, a type of green-manuring. No other type of fertilization such as the use of animal excreta was practiced.

In the grading and building of terraces for growing "water" or "wet" taro, *kalo wai,* and the construction of a system of aqueducts, flumes, and ditches, *hā wai* and *'auwai,* to bring water from dammed streams or springs to these terraces, the Hawaiians showed greater engineering and building skill, ingenuity, industry, and planning and organizing ability than any other Polynesian people.

All the many tasks associated with the preparation of the soil, planting, and often, harvesting were performed with the use of a simple, crude tool, the *'ō'ō,* or digging stick, used to dig up the soil to make it fine; to make the holes and trenches for planting various crops; and finally, to dig up the root crops such as taro, sweet potatoes, and yams. The hands were used for weeding, clearing, raking, and shoveling; the feet, to shovel and push soil around maturing sweet potatoes and dry taro; to stamp or tramp the bottom soil of the *lo'i,* the flooded taro terraces; and to make mounds on which the taro cuttings, *huli,* or shoots, *pōhuli,* were planted in the taro patch. Specific tools are mentioned in the descriptions of the food plants that follow.

Weather wisdom was of major importance to the Hawaiian farmer, not only from season to season, but from day to day. Weather was associated with two seasons: *kau,* summer, which began in May and was dry and hot; and *ho'oilo,* winter, beginning in

October, when it became rainy and chilly. The seasons were further divided into moon months; and, as with many other native people, preparing the soil, planting, harvesting, and processing of plant materials were all performed according to the phases of the moon. Even the time of day was supposed to have an influence on such procedures: for example, the favored time to plant bananas was at noon, when the sucker cast no shadow but "rested" or "retired" within the plant; thus all the strength went into the trunk and fruit and the latter matured rapidly.

The story of the early Hawaiian farmer would not be complete without commenting on the land system. The method of holding land in ancient Hawai'i was, in European terms, feudalistic. Ramifications of this land system were extremely intricate; it suffices to say here that the supreme chief, *mō'ī,* of an island did not own the land but only "held" it, serving as a trustee for the nature gods Kāne and Lono, who made the land fruitful. The king, retaining some of the portion assigned to him for his own use, partitioned the remainder into districts under the supervision of his high chiefs, *ali'i.* The latter, in turn, divided their allotted portion among their own lesser warrior chiefs, *koa,* dependents, and headmen or land supervisors, *konohiki.* These retainers would, in turn, subdivide their lands among the commoners, *maka'āinana,* who were the farmers, and their families.

Although changes took place in the stewardship of lands at the higher levels of Hawaiian society, there was as a rule a permanency in the occupation by a farmer family of land held by successive overlords. These tenant farmers were not serfs in the same sense as the medieval European term, for they could move to the land of another overlord, *konohiki* or *ali'i,* if they had reason to do so. At the time of the Great Mahele, or division of land, in 1848, under the reign of Kauikeaoūli, Kamehameha III, these tenant farmers became owners of the land they had cultivated; the term *kuleana* in the old sense of the word, meaning a portion or share and also a man's rights, was applied to this piece of land.

Water, which was used to irrigate his staff of life, taro, symbolized bountifulness for the Hawaiian farmer; thus the word for water repeated, *waiwai,* was used to designate something of great value: property, ownership, prosperity. Because taro, primarily, and to some extent other food crops depended upon water for their growth and productivity, the fundamental conception of property and law was based upon water rights rather than upon land use and possession. Actually, there was no conception of ownership of land, as already noted, or of water, but only the *use* of land and water.

The building and maintenance of the flooded taro terraces *(lo'i)* and the irrigation ditch system *('auwai)* were communal undertakings; all taro farmers took part in this labor, and all shared in the use of the water, the amount available to each being in part determined by the amount of labor contributed. The water rights of farmers along the streams themselves were respected in that no waterway was permitted to divert more than half the flow from any stream; each farmer took only as much water as he needed and then closed his inlet so that the farmer below could take his share. Thus this way of life was one of taking one's share and looking after the rights of his neighbors as well, without greed or self-interest.

The complex governance of water usage; the intricate irrigation system; the rituals attendant to the construction of the whole system; and the spiritual connotations of life-giving water—all these were evidence for the ancient Hawaiian's belief that these were sacred.

FOOD PLANTS *(na mea ulu i 'ai)*

The food plants of old Hawai'i can be divided into three groups: those known as staple foods, the principal starchy foodstuffs; those of less importance but still of significance because of their nutritive value and because they added variety to the diet; and those known as famine food plants.

Among the first group the most important was taro, *kalo.* Sweet potato, *'uala;* breadfruit, *'ulu;* and yam, *uhi,* are also starchy food plants but were less important than taro in the amounts consumed as sources of carbohydrates.

Food plants such as banana, *mai'a;* sugarcane, *kō;* coconut, *niu;* candlenut, *kukui;* Polynesian arrowroot, *pia;* and ti, *kī,* belong to the second group: those plants whose primary value was to provide nutrients not present in the staple foods and to add variety to an otherwise bland diet.

It was primarily the wild plants in the forest that provided food during times of famine, although some of the less important of the second group of plants became important when the staple food crops failed.

As already indicated, the Polynesian explorers who discovered Hawai'i found no important food plants; they had some food plants with them, and settlers who followed them probably brought more. These early introductions, which have come to be known as "Polynesian introductions," are taro, sweet potatoes, breadfruit, yams, bananas, coconuts, sugarcane, and Polynesian arrowroot; but the most important was taro.

Kalo (Taro)

Several Hawaiian legends describe the origin of taro. In the most frequently told version, Wākea—by some considered actually to be Kāne, the procreator of both man and Nature; by others, a man rather than a god—took Papa as wife. From this union was born a daughter, Ho'oho-kui-kalani. This daughter became with child by Wākea; she bore a premature infant, who was named Hāloa-naka-lau-kapalili, literally "long-stalk-quaking-trembling-leaf," referring to the trembling taro leaf on its quivering stalk. This first-born child died and was buried by Wākea; from the child's body sprang a taro plant. A second child was born to this couple and was simply called Hāloa; he became the ancestor of all people. Thus to the Hawaiian planter, taro, the first-born, was genealogically superior to and more sacred *(kapu)* than man, the second-born.

The status with which taro was consequently endowed was not the cause but the result of the place of this very important plant in native life. Thus we can speak of the culture—the cultivation—of taro, and Hawaiian culture—their way of life—as being closely interwoven. Further indication of this close association with and reverence for taro by the ancient Hawaiians is that the terminology used for taro was adopted into that for the family; just one example of this is the term *'ohana,* the family or kin group. *'Ohana* is derived from the *'ohā* of taro, the offspring (sucker) attached to the mother corm (underground stem). Also because of its sacred status taro could be planted, harvested, and mashed into *poi* only by men. However, women knew how to make *poi* and did so when they were left alone, when no men were about.

Taro is not solely associated with the islands of the Pacific; it has been and is grown within a wide central belt encircling the globe. It is cultivated in small or large areas, and is eaten in some form or other in such seemingly unlikely places as the Azores in the Atlantic Ocean, Sri Lanka and Indonesia in the Indian Ocean, and in Central America and Africa. Taro is mentioned in Chinese books as early as 100 B.C., and is reported as having been one of the established food plants in Egypt at the beginning of the Christian era.

Although grown in the above-mentioned areas, it was in the islands of the Pacific that this plant attained its unique and distinctive character, and Hawai'i was where planters brought it to the highest state of culture and where it played such an important role in the diet of the natives.

The Polynesians probably brought not more than a dozen varieties of taro with them, yet three hundred varieties have been listed as growing in precontact days. One may well ask: why so many varieties and where did they come from? Locality, use, and method of growing were all factors in the selection of varieties. There is evidence that the ancient Hawaiians knew the process of cross-pollination, and in this manner were able to produce a certain number of hybrids. However, far more new varieties would have resulted from mutation or sporting, *kāhuli:* some of the shoots produced on an established variety were different from the parent plant. From among these mutants the Hawaiians would consciously select those that grew best under various conditions as well as those most desirable because of size, quality, color, and other characteristics.

Varieties were named for their leaf, petiole (leaf stem), or corm color, such as *lehua* for its reddish corm and bright red *piko* (spot indicating petiole attachment to leaf blade), suggestive of the crimson blossom *(lehua)* of the *'ōhi'a* tree; or the *uahi-a-Pele* (Pele's smoke) with its green leaf "brushed" with dark smoke; or for some outstanding characteristic, such as the variety *'apuwai* with its cup-shaped *('apu'apu)* leaf, which catches and holds water *(wai);* or for fish having similar coloration, such as the variety *kūmū,* with its brilliant red petiole, named for a red fish, *kūmū.* Varieties were also classified as being "soft" or "hard" taros, referring to the texture of the corm flesh. Although some varieties could be grown both "dry" and "wet," and all varieties could be grown dry, not all dry varieties could be cultivated as wet taro.

Taro is a marsh plant; thus, when the Polynesians arrived and established their first villages along the seacoast, near the mouths of streams, it was natural that the first taro plantings were made in the swampy lands found there. With an increase in the established population and additional migrations from the central Pacific, the need to produce more food arose. Hawaiians then moved into the valleys, clearing the land of native vegetation. They widened areas beside the streams and springs to create the forerunners of the taro patches. Later, as more taro was needed, the Hawaiians developed an elaborate system of growing this plant in the flooded, banked, and terraced plots called *lo'i.* Taro grown in this manner was called "wet" or "water" taro *(kalo wai)* —not wetland taro, as it is erroneously called today.

Taro was grown in soil, as well as in flooded paddies; this was dry taro, *kalo malo'o,* and should not be called "dryland" taro, as it often is today, because the term *dryland* farming has an entirely different connotation on the mainland of the United States. Such taro plantings were made on the drier, lower slopes of mountainsides (known as *kula*) just below the forest limits, and in forest clearings.

A typical method of making a *lo'i* was as follows. The selected portion of land in which a new *lo'i* was to be made was first flooded for several days to soften the soil if rains had not already done so. The dimensions and shape of the *lo'i* were determined by the contour of the land. The soil within the designated boundaries was loosened with digging sticks and thrown up along the line of the proposed embankment. Digging continued until the desired excavation had been made and the subsoil layer reached. Banks were firmed by beating *(ku'i)* sugarcane leaves into the surface of the inner faces with a log *(lā'au)* or with the butt of a coconut frond *(lau niu).* Coconut fronds were laid on the top of the banks and pounded in with large, flat stones. Over the fronds was spread moist soil, which was then pounded until the surface was smooth *(ku'ikē).* Finally the banks were covered with fine soil and then with a mulch of leaves to prevent the banks from drying out too fast and cracking. Sometimes the inner faces of the banks were reinforced by lining them with stones.

When there was a decided slope to the land, stone retaining walls were built to form terraces. The topsoil was dug down to subsoil along the lines of the proposed walls to establish a firm base. The excavated topsoil was thrown up on the slope above the excavation to provide fill for the terrace. Hawaiians were noted as dry-wall masons, using no mortar. They gathered rocks from the fields, from streams, or from the beaches for the terrace walls. To make the walls more stable, they were sloped slightly inward toward the top.

After the *lo'i* had been excavated to the desired depth, just enough water was admitted to cover the floor several inches, and then it was puddled or treaded *(hehi)* to make the bottom watertight. Several days passed; more water was added and puddled a second time before planting in the bottom mud.

The propagating material most frequently used was a cutting *(huli)*. When the taro plant was harvested, the crown of the underground stem, along with the leaves, was severed with a special wooden cutter *(pālau)*. The leaf stalks were then cut to a height just above the groove in the last opened leaf so that the last-formed leaf, still furled within this groove, remained undamaged. Then the *huli* were tied in bundles and laid in the shade, often covered with grass, for a week or ten days, during which time a healthy crop of roots emerged. Although the method of planting *huli* in lines, prevalent today, is sometimes thought to have been brought to Hawai'i by Chinese immigrants it is likely that this was one of the methods practiced by the ancient Hawaiians. The other method consisted of forming mounds or ridges on the tops of which *huli* were planted, above the water level of the *lo'i*.

When the line method was used, rooted *huli* were stuck upright in the soft mud at the bottom of the *lo'i*, along a line probably made from *olonā* cordage; this was stretched from one side to the other of the patch. The mud was pressed lightly around the base of the *huli*, and the *lo'i* was then flooded. When the first two leaves had unfurled, the patch was drained and left so for two weeks, during which time the *lo'i* was kept damp either by rains or light irrigation. During this interval the roots elongated and became established in the soil. Then the *lo'i* was flooded again and remained so until water was withdrawn just before harvest. It is very important for taro to be irrigated with fresh, cool running water.

Buds on the corm develop suckers *('ohā)*, which were also used for propagation; these are smaller than the mother plant, with smaller corms of their own and fewer, smaller leaves but have the advantage over *huli* of possessing an established root system.

Seeds borne by the taro plant are fertile but produce weak plants; for this reason they were seldom used for planting. However, it is almost certain that a number of new varieties were obtained from seed plantings.

All parts of the taro plant were eaten, but the underground stem (corm) was most prized, for it was from this that the Hawaiians made their most important food, *poi*.

The unpeeled corms were steamed in an *imu* (underground oven). The cooked corm *(kalo pa'a)* was peeled with a large *'opihi* (limpet) shell, the back of which was filed off and sharpened on a stone; or with a sharp stone; or with a piece of shell. The peeled corms were first broken into pieces *(pākī'ai)*, then mashed *(ku'i)* with a stone pounder *(pōhaku ku'i 'ai)* on a heavy board slightly hollowed out on its upper surface, called a *papa ku'i 'ai*. Very little water, only enough to prevent sticking, was added to the taro as it was being pounded into a paste. The water was added by dipping the left hand into a calabash of water at the pounder's left side and passing the wetted hand under the mass of taro or, more frequently, over the lower face of the pounder.

The paste produced after proper pounding is stiff and was called *'ai pa'a* or *pa'i 'ai*. For journeys, this stiff paste was wrapped in large cylindrical packets *(holo 'ai)* protected by pandanus leaves *(lau hala)*. The thick paste could be kept for several months without spoiling.

To make *poi*, *pa'i 'ai* was mixed with water to a desirable consistency, this being primarily a matter of personal preference. The expressions "one-finger," "two-finger," and "three-finger" *poi,* designating *poi* that ranges from stiff to thin, are modern terms; in old Hawai'i to have eaten *poi* with three fingers would have been considered piggish! It was proper for women to eat with two fingers since they preferred a thin *poi* whereas men ate as they pleased, with either one or two fingers.

Hawaiians preferred fermented *poi;* fresh *poi* was reserved for babies and invalids. *Poi* was stored in covered calabashes to prevent a hard crust from forming on the surface and to keep out dust.

Portions of tender young taro leaves of all varieties or whole leaves of certain varieties were eaten as greens. The leaves were prepared for eating by wrapping them in ti-leaf packets called *laulau,* with fish, chicken, or pork, and steaming in an *imu.* Evidently both the uncooked and cooked leaves were called *lū'au.* In picking *lū'au,* the planter tore off only the upper half of an immature leaf, leaving the basal part of the blade to unfurl further and the leaf stalk intact to save the next leaf, growing furled within it.

Other parts of the taro plants eaten include the leaf stalk *(hā),* which was peeled and cooked as a vegetable, and the flower, which was especially relished because of its delicate flavor. Only the spadix was eaten; the spathe and its hard base were discarded. Flowers were cooked in the same manner as *lū'au* and *hā.*

A dessert called *kūlolo* was made from a mixture of grated raw taro, grated coconut flesh, coconut water, and the juice of sugarcane as a sweetener; this was wrapped in ti leaves and steamed in the *imu.*

Taro is very nutritious. The corm is an excellent carbohydrate food; the small starch grains are easily digested. It is a good source of calcium and phosphorus—ancient Hawaiians are noted for their sturdy bone structure and fine teeth. The corm also contains appreciable amounts of vitamins A and B. *Lū'au* is a source of vitamins A, B, and C; it is as good a source of minerals as spinach.

'Uala (Sweet potato)

As with taro, the Polynesians brought only a few varieties of sweet potatoes with them; in time these numbers were greatly increased (by one estimate, seventy or more) through natural hybridization or cross-pollination but primarily through sporting (mutations). Variety names were based on leaf and swollen root shape; the color of leaf veins; and skin, cortex, and flesh colors of the roots. To mention a few: the root of the *piko nui* variety is flattish and round like a swollen navel *(piko); uahi-a-Pele* (Pele's smoke) has a leaf that is dusky, as though brushed with smoke; and *pia,* which has the flesh color and consistency of the Polynesian arrowroot *(pia),* grown by the Hawaiians for its starch-filled underground tubers. Not infrequently the same variety was known by different names in various localities.

Sweet potatoes, although not as adaptable to variations in sunlight and moisture as taro, can be grown in much less favorable localities with respect to soil. This plant can adapt itself to any type of soil except clay, from rich earth made of decomposed lava and humus to volcanic ash and black sand; in the latter case humus, coral sand, and red earth were added.

The Hawaiians were adept in the selection of varieties best suited to various localities as well as to the season for planting. The time factor regulating the planting of sweet potatoes was one of variable cycles of weather rather than that of regular seasons. The general rule was that in dry localities slips were planted after rains had soaked the soil well several times; in damp areas the *'uala* planter waited until the rainy season had ended. It is maintained that, in general, sweet potatoes grow best in drier regions. Sweet potatoes were the staple food on Ni'ihau, the island of little water, where taro was not grown.

When growing sweet potatoes for the first time, a patch was prepared by burning off the grass and shrubs; the ground was then thoroughly dug with an

ʻōʻō (digging stick) and the stubble raked out with the fingers and thrown aside. Such a patch was called *māla ʻuala* and was used primarily for slow-growing varieties. Sweet potatoes could be grown in small pockets of semidisintegrated lava by fertilizing with decayed leaves or other organic material and piling fine gravel and stones around the vine stems. This type of cultivation produced inferior sweet potatoes, however. So far as is known, there was no systematic irrigation in sweet potato culture. However, when there was an abundance of water in summer, any water beyond that necessary for taro culture was diverted to raise a second crop of sweet potatoes. Also, in dry seasons, when the supply of water was insufficient for taro growing, *loʻi* were utilized to grow sweet potatoes, using whatever little water was available.

The propagation of sweet potatoes was only from cuttings or slips *(lau ʻuala).* Although it is thought that new varieties were obtained by planting seeds, this and the planting of "seed potatoes" were not usual methods of propagation. Hawaiians considered old vines a better source than long, luxuriantly growing runners for propagating material, because the latter produce many roots but very few, small sweet potatoes *(au).* About a foot of the tip end of a vine was broken off for the cutting; all leaves except three or four at the tip were removed.

In a dry area or during a dry season, the slips were collected in the evening, if possible after a shower, bundled up, and left on the ground to be planted the next morning. If the day promised to be a hot one cuttings were planted the next evening. However, if rain was expected the slips were bundled, covered with grass or other plant material, and placed in shade for two or three days, during which period rootlets *(kiʻu)* emerged from the nodes. Actually, bundles of prepared cuttings could be kept fresh and damp for planting for as long as a week by wrapping them in ti leaves and placing these bundles in shade.

As for all crops, there was a most-favored time for planting; for this plant it was at the beginning of the dark phase of the waxing moon. Although there were opinions of various planters that differed from this, all agreed that on certain days slips should *not* be planted (for example, in the dark of the moon).

Hawaiians usually planted sweet potatoes in mounds *(puʻe).* Where the soil was dry and powdery, the earth was heaped up into low mounds in no particular order or without any exactness. But in areas of high rainfall or in some continuously damp lowlands, symmetrically arranged high mounds were built to provide ample aeration for roots, a necessity in growing this crop. Slips, two to three to a mound, were placed vertically in holes made with a digging stick. The base of a cutting was inserted to a depth of about half a foot and the earth pressed down around it, usually with the feet. After planting, the mounds were covered with a mulch of leaves to conserve moisture, with care taken that the potato leaves were not covered.

Although usually only one variety was planted on a mound, to extend the harvest period planters frequently planted a quick-maturing variety along with a slow-growing one that could be left in the ground for many months after maturing.

As the cutting elongated into vines *(kā),* the soil around their roots was stirred with a stick on several successive days; in dry localities this was continued until a shower soaked the soil, after which the moist soil was reheaped around the vines. In all localities it was important to keep the soil piled around the base of the plant. This task of softening the soil, hilling, and pressing the earth down around the base of the vine was called *kaiue.* Weeding *(ʻōlaʻolaʻo)* was kept to a minimum; and once the vines covered the mounds there was little need to weed.

The rapidly elongating vines were confined to the mounds in one of two ways: each vine was twisted *(wili)* around its own base and covered with firmly pressed soil, or the vines were trained to follow around the slopes of the mound, pegged down with V-shaped pegs made by trimming small forked twigs. In this way, root formation was confined to the mounds.

In areas with frequent showers, or after a heavy rainfall, vines showing thick growth were turned back, with the underside exposed, so that wind and sun would dry the foliage and soil beneath; this prevented rot and mildew.

When sweet potatoes began to form, about a month or so after vines had covered the mound, a planter reached into the mound *(kilo ʻuala)* and broke off and removed the smaller potatoes *(hahae)* as a sort of pruning, while the larger ones were left undisturbed to mature to full size. After the smaller potatoes had been removed, a *kapu* was placed on the patch to ensure that the rest were left undisturbed. The small potatoes were offered at a sacrificial feast where prayers to the god of sweet potatoes, Kamapuaʻa, for full maturity of the remaining potatoes were said; after this ceremony no one was allowed to enter the area.

When the remaining sweet potatoes began to mature, only enough of the largest potatoes to fill current needs were dug at a time, usually with an inthrust hand or short stick, care being taken not to disturb either the vine or the remaining crop.

After the harvest of the crop, old roots and unharvestable potatoes remained in the ground to furnish slips or cuttings for the next planting.

The worst enemy of sweet potatoes in old Hawaiʻi seems to have been the rat. There was also a dry rot of the potatoes caused by excessive rain and humidity combined with heat or by negligence or poor cultivation.

Sweet potatoes supplied the diet of the ancient Hawaiians an additional important carbohydrate. They were cleaned and placed in an *imu,* unpeeled, for roasting. They were served whole and then broken into pieces for eating away from the skin.

Hawaiians also made a *poi* from baked, peeled, and mashed potatoes; although they regarded this *poi* as dietetically superior to taro *poi* it was considered less satisfactory as a staple food because it fermented very rapidly.

A kind of pudding, *piele ʻuala,* was made by mashing cooked and peeled potatoes, mixing with coconut cream, wrapping the mixture in ti leaves, and steaming the packet in an *imu.*

The young leaves were used for greens *(palula).* The leaves of all varieties were edible but those from the small-leaved varieties were the most delicate and therefore most flavorful.

The ancient Hawaiians made use of their knowledge of the fermentation process and the rapid fermentation of sweet potatoes to produce a beer, *ʻuala ʻawaʻawa.* Perhaps they discovered this beer by accident when their sweet-potato *poi* was left too long!

ʻUlu (Breadfruit)

An old Hawaiian myth describes the origin of breadfruit as follows: The god Kū, to save his children from starvation during a time of famine, buried himself alive in the earth near his home. Kū had told his wife that a tree bearing fruit shaped like a man's head would spring from his body: "My body will be the trunk and branches; my hands will be the leaves; the heart, core, inside the fruit will be my tongue. Roast the fruit, soak it, beat off the skin, eat some yourself, and feed the children." This myth, considerably shortened here, undoubtedly originated in Tahiti where breadfruit played a much more important role as a staple food than in Hawaiʻi. The locale of the Hawaiian version of this myth is supposed to have been at Kaʻawaloa in Kona, on the southwestern coast of the island of Hawaiʻi.

Although there are many varieties of breadfruit in other parts of Polynesia, there was but a single one in old Hawaiʻi. The lack of development of more varieties seems to indicate that it was introduced considerably later than the other staple food plants, taro and sweet potato. This might also explain why there were not more extensive plantings here. This was unfortunate because its presence would have been a great asset during famines.

The Hawaiian variety of *'ulu* is without seeds; it was propagated from root suckers *(weli* or *hehu),* a shoot growing from a lateral root. Such suckers often originate spontaneously or may be initiated if such a root is injured by deliberately "nicking" it. Usually a lateral parent root was severed on both sides of the sucker; this induced the sucker to form its own roots. When the latter had grown sufficiently the rooted sucker was dug out with a ball of adhering soil and transplanted into a hole just large enough to receive the ball of soil.

Breadfruit trees were usually planted in more or less extensive groves along the southern coasts of the Islands. To a lesser degree, they were grown inland along windward coasts and up valleys in sheltered places.

The fruit was harvested with special pickers called *lou* and then prepared for eating in several ways; the simplest was to broil it in its skin on hot embers. When thus prepared it was called *'ulu pūlehu.* The fruit was also made into *poi 'ulu;* to prepare this the fruit was baked in an *imu,* peeled, "cored," and the flesh then mashed with a little water. Unlike *poi* made from sweet potato, this *poi* kept as long as that made from taro. However, Hawaiians considered it inferior in taste and nutritive value to *poi* made from taro, and also less desirable because it produces much gas in the stomach.

Perhaps the most relished food prepared from breadfruit was the dessert *piele 'ulu.* This was made by mashing the flesh of very ripe fruit and mixing this with just enough coconut cream to make a stiff paste. This paste was wrapped in ti leaves and baked in an *imu. Pepeie'e 'ulu* was also made from the flesh of very ripe fruit but with large quantities of coconut cream. This was also baked in an *imu;* after it had cooled this pudding was sliced and dried in the sun. This caused a hard oily film to form over the surface of the slices. Placed occasionally in the sun to prevent mildewing, this product would keep from the end of one bearing season to the beginning of the next.

Uhi (Yam)

In old Hawai'i there were three species of *Dioscorea* included under the general name *uhi.* The only one cultivated, *D. alata,* was called *uhi;* the other two, *D. pentaphylla,* known as *pi'a,* and *D. bulbifera,* called *hoi,* grew wild. Of these two latter species, *pi'a* was utilized as an emergency food while *hoi,* which was poisonous, was not eaten unless it was prepared in a special manner described later. All three yams were Polynesian-introduced.

The Polynesian settlers probably brought several varieties of the edible yam with them to Hawai'i. These numbers were increased through mutations and perhaps even through cross-pollination. Two varieties, *ke'oke'o* and *'ula'ula,* were universally grown throughout the Islands; the other varieties were limited to one island or to a specific area. *Ke'oke'o,* as the name indicates, had a tuber with white flesh, while the red skin of *'ula'ula* gave that yam its name.

Uhi was propagated from small "seed tubers," seedling or baby (secondary) yams *(hua uhi)* that form in addition to the main tuber; or from buds sprouting from "eyes" *(maka)* in pieces of the top of the mature tuber. In most planting systems both types of planting material were set in holes in the ground to a depth of about a foot, then covered with lightly packed soil. As the vine grew, soil was heaped up around the base of the stem, which, in turn, was covered with leaves and/or other plant debris as a mulch.

Yams were grown extensively on all the Islands; in general they were planted in inland gulches, on high semiforested *kula* slopes, or in the lower zones of rain forests. In all these locations yams were planted adjacent to dead trees, or those that had been girdled to kill them, to provide support for the vines.

There were two other methods of planting. One

was in large bins made of horizontally laid trunks of tree ferns *(hāpuʻu)* made like the walls of a log cabin; these were built directly on the ground. The bin was filled with decaying fern fronds and other plant debris. Seedling tubers were thrust into the debris a few inches below the surface. No soil was added to the bin but as the debris became further decayed and sank, more organic material was heaped on the top of the bed.

The third method of planting was practiced on steep hillsides and on the banks of gulches. A vertical hole was dug in the side of the slope, and a large flat stone was placed in the bottom. The hole was then filled with soil and decaying leaves and the seed yam planted shallowly, near the top of the hole. The stone at the bottom of the hole prevented the tuber, which grows downward, from extending too deeply into the ground and forced the root system to spread out.

Toward the end of the rainy season, in February or March, the planted seed yam, or the tubers that had remained in the ground, began to sprout. By late spring or early summer the vines had elongated so that they twined around the branches of trees for support if this was the planting method, or the vines were festooned over the sides of the *hāpuʻu* bin if planted there, or over the surrounding ground if planted in a slope-hole.

The foliage matures in October and withers in December. Withering indicates that the plant is mature, but the tubers continue to fill out. For this reason the tubers were left in the ground during this period of foliage dormancy. The tubers were dug up after the rains had started again. By that time the tubers had reached their maximum size and were succulent. If the tubers were left in the ground until the end of the dry season they became mealy *(punapuna)* and were therefore less desirable; if left in the ground until the new vine had grown the old tuber became watery *(loliloli).*

Yams are generally not considered to be an important staple food of the ancient Hawaiians, one reason being that *uhi* could not be made into *poi.* However, on Niʻihau, where little taro was grown because of lack of an adequate water supply, yams, along with sweet potatoes, were definitely a staple food plant.

Two other yams were mentioned earlier in the chapter. *Piʻa* is edible but was far less important than *uhi* and was probably never cultivated. It grew wild in forests, where it climbed up trees for support. The ground tubers are edible when cooked but were usually used only as a famine food. They were eaten warm; cold, they were considered fit only for pigs. *Hoi* is commonly called the poisonous or bitter yam. As with *piʻa,* this yam grew wild in the forest and was never cultivated. It, too, grew on trees. As its common name indicates, the flesh of both the underground and aerial tubers is poisonous. However, if properly prepared, boiled in lime and steeped in water for several days with frequent changes of water, the underground tuber can be eaten, and probably was by ancient Hawaiians during a famine.

Maiʻa (Banana)

Maiʻa was not a staple food of the ancient Hawaiians; rather this fruit furnished a pleasing variety to their diet. However, bananas gained considerable importance in times of famine. Because of its lesser importance as a food, there apparently was no systematic cultivation in extensive plantings as with taro and sweet potatoes. Instead, clumps of bananas were planted near their dwellings and on the banks of taro terraces. *Maiʻa* were also planted in such moist and sheltered areas as little gulches on the lower fringes of the forest and in the median forest belt. Protection was needed against high winds that tore leaves and caused the plants to topple since banana has no true trunk. These forest and forest-fringe plantings served as insurance against famine.

When the mother stalk was cut down with the

harvest of the fruit, this stimulated the growth of suckers initiated from the buds on the underground stem *(mole)*. These suckers grew to form a clump or cluster of plants called *ōpū mai'a*. Others were removed for propagating material; these latter were carefully cut away with part of the corm. A hole was dug, the sucker was set in the bottom, and soil was added. Generally, the planted banana was not mulched, as many plants were, because the banana sucker was placed so deep.

The large number of varieties in ancient Hawai'i indicates that bananas were brought here at an early stage of settlement. Varieties were divided into three general classes: those eaten raw; those eaten either raw or cooked; and those eaten only after cooking.

A few examples of those eaten raw were *mai'a hāpai* (pregnant banana), one that was unusual in that the ten or less small, sweet yellow fruit develop and mature within the trunk; *mai'a pōpō 'ulu* ("breadfruit-like ball-shaped" banana); and *mai'a māhoe* ("twin" banana), which has a branched fruit stalk bearing two bunches of fruit that are small and yellow-skinned and have flesh that is light salmon in color.

Among the varieties eaten either raw or cooked were *mai'a hilahila* ("bashful" banana), which has a green, purple, and pink trunk, and fruit whose flesh is salmon pink; *mai'a ha'a* ("low" banana), with a short trunk and yellow fruit; and *mai'a koa'e,* with beautiful striped leaves, trunk, and young fruit, and fruit that is round and yellow.

Included in the varieties that were eaten only after cooking (the greatest number of varieties belong to this class) were *mai'a eka* ("discolored" banana), the skin of which changes from red to green to yellow as it matures; and *mai'a-ka-ua-lau* ("many raindrops" banana), so-called because the skin of the dark-green fruit has light-green spots like raindrops; when ripe, the skin becomes yellow and has a waxy appearance; the flesh is a light yellow. *Mai'a puhi* ("eel" banana) is another variety cooked before eating; when young the fruit is twisted like an eel, hence its name. When mature the fruit is long and "fat," the skin medium thick and yellow, and the flesh yellow.

All varieties except three were *kapu* (forbidden) to women.

Bananas grown in the uplands were ususally ripened in pits; bunches of mature fruits were wrapped in dry banana leaves in pits also lined with dry banana leaves to ripen. Fruits so ripened were considered better flavored, sweeter, and more fragrant than those ripened on the plant as in the lowlands.

Those varieties that needed to be cooked before eating were steamed in an *imu* and then eaten out of their skins. *Mai'a pūlehu* were bananas roasted over embers or in hot ashes. In time of taro shortage a *poi* was made by mashing ripe fruits and allowing this "mash" to ferment. Certain sweet varieties were eaten as *pūpū* by *'awa* drinkers to "top off" the drink and take away the bitter taste; this pratice is the origin of the term *pūpū* applied to the hors d'oeuvre served with alcoholic drinks in today's society.

Niu (Coconut)

Niu was very important to the ancient Hawaiian because of the many uses they had for various parts of this palm. Its use for food, however, was minor.

The liquid endosperm of the fruit which occupies the central cavity of the nut is called coconut water *(wai niu* or *wai o ka niu)* and was drunk on journeys where fresh water was not available. The solid endosperm, the so-called meat or flesh, was grated *(olo)* with a special grater *(wa'u niu)* made from a prepared shell (see section on food utensils). The grated flesh was then squeezed in a piece of the fibrous sheath of a young coconut frond; when the surface of this is scraped to remove debris, the sheath has the appearance of cheesecloth which, today, we often use for the same purpose. The cream (*wai o ka niu* and *wai niu* are used to designate coconut cream as well as coconut water) was used to make puddings, such as *haupia.*

Kulolo was made by mixing grated raw taro and grated coconut meat in equal quantities with a quantity of coconut water and the raw sap from a length of sugarcane stalk as sweetener. This mixture, rather thick, was wrapped in ti leaves and baked in an *imu.* When *kulolo* is made today, *poi* and coconut water or cream are substituted for taro and coconut flesh, and honey or crystalized brown sugar is added as the sweetener.

Haupia was made from coconut cream thickend with *pia* (Polynesian arrowroot). The prepared *pia* was added to the coconut cream in a quantity required to thicken the cream upon heating. This mixture was placed in a package *(pū'olo)* made of closely overlapping fresh ti leaves to prevent leakage of this liquid. The making of such a package was probably facilitated by placing the leaves within a gourd or coconut shell bowl. The ends of the leaves, protruding beyond the rim of the bowl, would be gathered together and tied with a dry half ti leaf to complete the *pū'olo* that was then placed in an *imu* to bake. A solidified pudding resulted which was cooled and then eaten out of the hand. Today, corn starch is used instead of *pia* for the thickening agent, and the pudding is cooked on a modern stove.

Kō (Sugarcane)

That the ancient Hawaiians had a great number of varieties of sugarcane, adaptable to many localities, implies that this plant was an early Polynesian introduction; and the existence of several groups of this plant, with quite different characteristics, indicates that more than one variety was originally brought here. Certain sugarcanes and fishes having corresponding colors had the same names.

Although sugarcane forms seeds, these were not used for propagation; instead, sections of the stalk were used, as today. Hawaiians considered the selection of vigorous stock all-important; the upper part of the stalk fulfilled this requirement. Terminal leaves were cut off and the stalk cut into two or three sections, each having two or three nodes. These sections, known as "planting pieces" or "seed pieces," were called *pulapula,* a term still used on sugar plantations in Hawai'i today. A trench was dug, the soil at the bottom was softened, and the *pulapula* was laid flat, horizontally, on the soft earth at the bottom of the trench and covered with soil. Grass or leaves were placed on top as mulch.

When the sprouts arising from the several buds at the two or three nodes appeared above ground, the sprouts were covered with soil and mulched. At intervals during the further growth of the cane earth was softened and heaped up to make a mound *(pu'u lepo)* around the base of the resulting clump.

Kō was planted near houses if it was not too dry there so that it was readily available for children to cut and chew. It was also planted along the embankments of *lo'i* in wet-taro culture. In dry-taro and sweet-potato fields on *kula* slopes, or in the lower forest zones, sugarcane was planted as hedges along lines of stones and rubbish thrown up between these fields. Thus the Hawaiian farmer utilized otherwise wasted space and provided windbreaks for the more delicate taro and sweet potatoes. And, of course, these hedge plantings provided an additional source of food.

The stalks were normally cut before tasseling, about a year after planting, to retain the highest sugar content. The upper portion was saved for propagating material; the remaining portion was stripped of its leaves and cut into sections. After peeling off the hard "rind," the pulp was pounded to soften and break the fibers, and the juice was squeezed out with one or both hands.

The fresh undiluted juice was used as sweetening for puddings made from taro, sweet potatoes, breadfruit, and bananas, described elsewhere. Toasted over an open fire, it was fed to nursing babies. Children chewed the whole "peeled" sections between meals as a sweet; they were encouraged to do this because parents felt that this strengthened the children's gums and teeth.

Kī (Ti)

The natural habitat of the ti plant was in the wetter open forests at lower elevations, where it grew wild. It was also cultivated near buildings and was frequently grown on the banks of taro patches. Ti is easily propagated from sections of the stem; these were simply thrust into the ground where desired.

It was the root of this plant that was used for food; this contains the reducing sugar fructose, which when steamed in an *imu* turned a molasses brown as a result of caramelization of the sugar. Sections of the cooked root were chewed by both children and adults for a favorite sweet. However, when eaten in quantity, this sweet produced diarrhea.

Ti leaves were used in the preparation, transportation, and storage of food; these will be described in the appropriate sections following.

Pia (Polynesian arrowroot)

Although the tuber of *pia* is rich in starch it was not a staple food plant for the ancient Hawaiians; in itself it was not an important food plant except in times of famine. In ordinary times, as a food, it was used primarily as a thickening for puddings.

Pia was planted in those areas where dry taro was grown; on the banks of *lo'i;* and around houses when and where rains were ample. Although the plant produces seeds, it was the tuber that was planted. At the end of the rainy season a hole was made in the ground with a digging stick, the tuber was placed in it, and then covered with a little soil.

The starch, which is bitter when raw, was prepared for use by grating the flesh and soaking the starchy pulp in a calabash with fresh water. When this material settled to the bottom, the water was poured off and fresh water was again added. This process of decanting and adding fresh water was repeated until the undesirable bitterness was removed from the starch. This process was called *hulialana* by the Hawaiians. After the bitter element was removed, the slurry was strained to separate the refuse from the starch suspension. On standing, the starch settled and the water was drained off. The thick paste was then spread on a flat stone to dry; when dry it was scraped off the stone and pounded into a fine powder. It was used as a thickener for a pudding, *haupia* (see coconut).

Wild Food Plants *(lā'au makanahele)*

Although the wild plants of the forests and upland gulches, and even those in the lowlands, were occasionally used for food at all times, it was during famines that these plants, discussed following, gained considerable importance.

FERNS. Probably the most important of these were the so-called tree ferns, *hāpu'u pulu* and *hāpu'u 'i'i.* It was the cooked starch that was used during famines. The top section of the trunk, with fronds, was cut off and the outer sheath, consisting of layers of fibers and "wood"—a very hard layer surrounding the starch core—was chipped away to expose the starchy pith. The latter was wrapped in ti leaves and baked in the *imu.*

Other ferns also provided important famine foods. Among these were *'ama'u,* the starchy pith of which was cooked in an *imu,* and the young shoots eaten raw or cooked.

Additional ferns used as food were the now-rare *pala,* whose frond stems had thick starchy bases that were eaten either raw or cooked; the high-altitude *hō'i'o,* whose young leaf shoots *(pepe'e)* were eaten raw with mountain-stream shrimps; and *kikawaiō* or *pakikawaiō,* whose roots and young fronds were eaten raw; they were also grated and salted to taste to make a slimy preparation somewhat like okra.

Roots and tubers. Roots and tubers of other plants were also eaten in times of famine, such as *ʻuala koali,* sweet-potato morning glory; *pi ʻa,* the wild yam already described; and *ʻape.*

Berries. There were a few native wild berries that were eaten for refreshment on journeys but became more or less important as food during famines. These berries included *ʻōhelo, pōpolo,* the coastal or beach *naupaka kahakai,* and *ʻūlei.*

Fruit. Fruits of *lama* are pleasantly sweet. Those of *ʻōhia ʻai,* the mountain apple, are palatable but if eaten fresh in any quantity cause stomachache. Therefore, Hawaiians split the fruit and strung the halves on bamboo splinters or the midribs of coconut leaflets. So prepared the fruit was allowed to partly dry in the sun to become *ho ʻoma ʻema ʻe* or simply *ma ʻema ʻe* (cleaned). The immature fruit of Indian mulberry, *noni,* edible if not very palatable, became by necessity a food during famines. Another was *ʻākala,* meaning pink, named for the color of the juice of two species of Hawaiian raspberry.

Leaves. Leaves only or whole young plants of several species were cooked in times of food scarcity. Among these were leaves of *pōpolo,* already mentioned because of its edible berries; *ʻāheahea,* a native pigweed or lamb's quarters; and a native lobelia, *ʻakū ʻakū.* The leaves of the three species were cooked in an *imu* wrapped in ti leaves.

In addition to the plants listed in this section, wild varieties of otherwise cultivated food plants growing in upland valleys and forests, such as taro, bananas, yams, and others, were also gathered during times of a shortage of regular, cutivated varieties.

Relishes or Condiments *(mea hō ʻono ʻono)*

The diet of ancient Hawaiians was quite bland. They did use a considerable amount of salt to season their food, but they used no plants that we consider spices. They did however use two plants as *mea hō ʻono ʻono* (literally, a tasty food). These were *limu,* both fresh-water and marine algae, and the kernel of the *kukui* nut.

The numerous edible seaweeds of old Hawaiʻi varied greatly as to locality, season, and regional nomenclature, making it impossible to name and describe them all here. One that was especially popular was *limu kohu,* a soft, succulent seaweed with a tufted appearance. It suffices to say that any of the *limu,* primarily the marine forms, furnished a spicy addition to the Hawaiians' food.

Men in canoes gathered those *limu* growing on rocks in rough water, away from shore. The marine algae that broke loose from offshore rocks on which they were fastened and floated in the sea near shore are called *limu lana.* These were gathered by women who waded out to collect them. Sitting on shore, they cleaned the *limu* by removing small crustaceans and sand. If the *limu* needed to be kept for a time, it was placed in salt water to prevent "wilting." To serve, the seaweed was chopped and placed in separate vessels and eaten with fish and *poi.*

Limu growing on rocks in streams were gathered by women; these required little cleaning. They were prepared for eating in the same manner as the marine seaweeds. The freshwater algae were not as flavorful as the marine species because they did not have the spiciness of the latter. They did, however, provide additional essential green vegetable food.

Although the raw *kukui* kernel has a cathartic effect, it became a prized relish *(ʻinamona)* when roasted and mashed with salt.

FOOD PREPARATION

(ka ho ʻomākaukau ʻana o ka ʻai)

Cooking *(ho ʻomo ʻa mea ʻai)*

Some plant foods were eaten raw but most were prepared before eating by one or more of the following methods: *kālua,* a combination of roasting or baking and steaming in an *imu; hākui,* boiling or steaming; and *pūlehu,* broiling.

ROASTING *(kālua)*. An *imu* was a shallow hole in the ground. In good weather, cooking was done in an outdoor *imu*. For rainy weather, the *imu* was sheltered by a shed *(hale imu)* with a roof of thatching and sides walled with stone. Sometimes the shed consisted of just the thatched roof set on posts with those sides facing the prevailing winds protected with plaited coconut fronds or bark. There had to be two *imu* for each family because the food for women had to be prepared separately from food prepared for men.

The *imu* and its contents were prepared as follows: kindling wood *(pulupulu)* was placed in the middle of the hollow, with large pieces of firewood *(wahie)*, branches and trunks of trees, laid around and over the kindling; a layer of specially selected porous lava stones, roundish stream or beach boulders about the size of a man's closed fist called *'eho* or *pōhaku imu*, was arranged over the wood.

The kindling was lighted by means of a fire-plow *('aunaki)*. A roughly pointed piece of very hard dry wood, called *'au lima*, was held in a forward slant, the fingers of both hands clasped over the front and the thumbs to the back. This handpiece was rubbed or pushed back and forth, very rapidly, against the *'aunaki*, a piece of soft wood such as *hau*, which was held in place between the feet of the firemaker, or those of an assistant, or with the front end pushed against a fixed object. The rubbing action of the hard handpiece against the soft wood of the footpiece produced a groove. The wood particles and dust resulting from continued rubbing were pushed to the forward end of the groove and when sufficient wood dust and particles had accumulated, the rubbing action was quickened until friction created enough heat to cause the material to smolder and smoke. Some flammable material such as dry old tapa, usually plaited into a loose three-ply braid with one end frayed for easy ignition; coconut fibers; or coconut stipules was placed on the ground and the footpiece emptied of its smoldering contents onto this pile. The overturned *'aunaki* was given a sharp tap on its bottom with the handpiece to dislodge all its contents. Sometimes the footpiece itself, if it was dry enough, became ignited and was used as kindling. The fire was fanned into flames with a bamboo fire-blowing tube called *'ohe puhi ahi*.

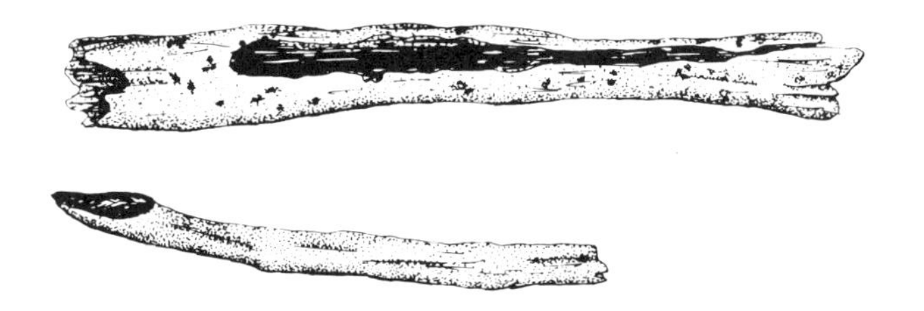

'Aunaki with *'au lima*

When the fire in the *imu* had burned out, the embers and any unburned wood were prodded out with a stick and the stones leveled to make an even floor. Over the hot stones was spread a layer of plant material: ti or banana leaves; *kūkaepua'a*, a small native creeping weedy grass; *'ilima kū kula*, a wild form of the usual *'ilima;* and/or seaweed to prevent the food from being scorched. On this protective layer were placed the plant products to be cooked: unpeeled and unwrapped taro corms, and ti-leaf-wrapped sweet potatoes, breadfruit, and various "greens" already mentioned, along with pig, dog, and chicken. Frequently added to the above-mentioned contents were split trunks of banana to create more steam, and leaves of a mountain ginger to give the roasting food a pleasing flavor and aroma. When the *imu* had been properly "loaded," the contents were covered with layers of leaves, preferably ti leaves, and an outside layer of *lau hala* mats was placed as the uppermost cover. Apparently a final cover of earth, used in some other parts of Polynesia, and now used in Hawai'i, was not used in old Hawai'i.

The length of time for proper cooking of food depended upon the size of the hole and the items cooked; sweet potatoes took about two hours, taro two to six, and the meats even longer. When the *kālua* process was complete, the coverings of the *imu* were removed and the contents placed in appropriate receptacles. Although other Polynesians used various contrivances for removing the hot food from the oven, Hawaiians picked it up with their bare hands, often dipping these in a calabash of cold water from time to time.

Boiling or steaming *(hākui).* Hawaiians lacked fireproof vessels in which to boil food; this was therefore placed in calabashes with water into which red-hot stones were dropped. The process was more one of steaming because only a small amount of water was used. It was by this method that many greens were cooked: young taro leaves and the tender ends of sweet-potato vines. As already noted, these greens were also cooked in *laulau* in the *imu.*

Broiling *(pūlehu).* This method of cooking was usually used in preparing food out in the field, or when only a short cooking period was required as for fish. Broiling was done over hot embers or ashes, often with a circle of stones confining it. Breadfruit and unripe bananas, unwrapped but in their skins to protect the fruit-flesh from scorching, were placed directly on the embers; greens and fish were wrapped in ti leaves first.

Equipment *(ka lako o ka ho'omākaukau)*

Ancient Hawaiians used a variety of implements in food preparation. The main classification of these implements, along with brief descriptions, follow.

Scrapers *(wa'u).* *'Opihi* (limpet) shells with their naturally serrated edges were used to peel taro and sweet potatoes. *Wa'u niu,* cut-down large cone or cowrie shells, with one end serrated, were used to grate the flesh of coconuts.

Knives and cutters *(pahi).* These were made from stone, bamboo, and sharks' teeth. By cutting one end of a strip of bamboo diagonally, the resulting serrated edge made a good cutting surface.

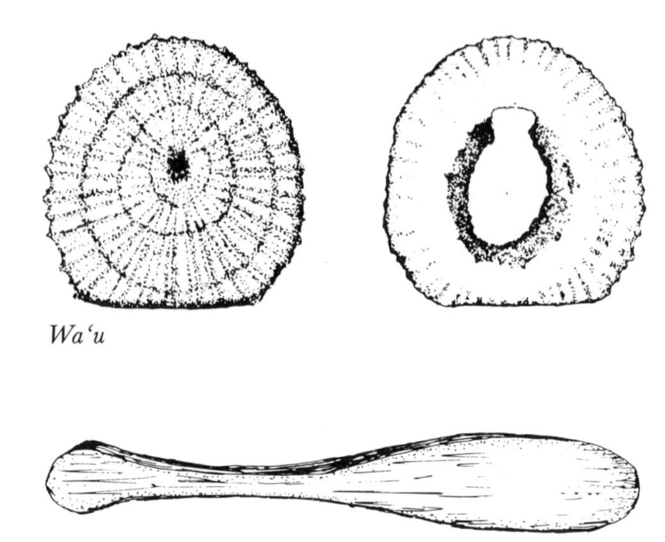

Wa'u

Lā'au ho'owali 'ai

Breadfruit splitters *(pōhaku kōhi).* Although in various parts of Polynesia wooden tools were used to split breadfruit, Hawaiians usually made theirs out of stone, as the name indicates.

Sweet-potato poi mixers *(lā'au ho'owali 'ai).* Unfermented sweet-potato *poi* was a favorite dish of the Hawaiians. A wooden instrument consisting of a handle and spatulate blade was used to mix the mashed sweet potato with water to form a kind of gruel.

Poi-pounding boards *(papa ku'i 'ai).* Boards on which cooked and peeled taro was pounded *(ku'i)* to make *poi* were usually made from breadfruit trunks. They were flat, shallow trays somewhat rectangular in shape with corners rounded off. In older boards, the outer slope was rounded off to meet the flat bottom without an edge. The rim along the sides and ends was either rounded or flat. The inner slopes of the sides and ends met the flat internal surface at an angle; bottoms were thick.

There were two classes of boards: (1) a short

Papa ku'i 'ai with *pōhaku ku'i 'ai (right)* and *pōhaku ku'i 'ai puka (left)*

board at which only one man worked and (2) a long board for two men, one at each end.

STONE POI POUNDER *(pōhaku ku'i 'ai* or *pōhaku ku'i poi).* Although this book deals with plant-derived artifacts, the poi pounder is discussed here because of its importance in making the Hawaiians' staple food. There were three general types, all made of porous lava rock. The form used on all the islands except Kaua'i had a neck with a terminal knob to prevent the hand holding it from slipping off, and the main portion, the pounding body, flared outward and downward from the neck to the actual beveled pounding surface. Ones used on Kaua'i were the ring pounder *(pōhaku ku'i 'ai puka),* which had a ring through which the thumbs of both hands were passed to grip the sides of the arch; and the stirrup pounder, shaped somewhat like a saddle stirrup, with the surface of the front side ground out with the oppposite side straight or sloping. This last beater was grasped with both hands from the back, the fingers passing around the sides of the hollow in the front and the two thumbs passing vertically over the top edge.

FOOD AND DRINKING UTENSILS
(ka lako mea 'ai a me ka lako mea inu)

A variety of objects made from gourds, wood, and other plant material filled the ancient Hawaiians' need for containers for eating and drinking and food storage.

The term *calabash,* originally applied to a bowl made from the gourd fruit, now is the generic name for bowls made from both gourds and wood. In Polynesia, at least, calabash has also come to include bowls made from coconut shells.

Gourd Utensils

Gourds, brought to Hawaii by the early Polynesians, were used to a greater extent here than anywhere else in Polynesia. One of the important uses of gourds was the making of food bowls and water bottles.

Two general types of gourds were grown: the poisonous, "bitter" gourd, *ipu 'awa'awa,* and the nonpoisonous "sweet" gourd, *ipu mānalo.* A further classification within the latter indicates that there were many varieties, differing in size and shape; those used for food containers were called *ipu 'ai,* those as water bottles, *ipu wai* or *hue wai.* Although the size of gourds used for food bowls varied greatly, only two shapes were used, a squat form with the diameter greater than the height, and a deep one with the height greater than the diameter.

Gourds were cultivated where they grow best: on the hot shores and in the lowlands, on leeward and southerly coasts of these Islands, where there is moderate rainfall and abundant sunshine. A great deal of attention was given to the growing of this plant. The seeds were planted at the beginning of the rainy season; the fruit is ready to harvest in six months, so the gourds had the hot, dry summer in which to mature. It has not been verified that lashed bamboo trellises were constructed to support growing vines. The usual account is that as the fruit increased in size a small prop or frame was made of three small branches, between which the gourd hung suspended to ensure symmetrical fruit. The ground beneath the suspended gourd was cleared of stones and pebbles, and a layer of grass or leaves was spread to prevent marring the undersurface of the fruit. Sometimes a ring of twisted grass

'Umeke pōhue or *ipu pōhue*

'Umeke pāwehe

was made for the gourd to rest on, for the same purpose.

Gourd bowls (*'umeke pōhue* or *ipu pōhue*). Gourds were picked when the fruit stem withered. Gourds for food bowls were processed in the following manner: The top (the stem end) was cut off and the gourd filled with seawater. This water was changed every ten days, thereby eliminating the "acidity" and softening the tough flesh. Another method was to remove the pulp and seeds after the top had been cut off, and then, when the rind had dried, to rub the inside with coral of successively decreasing roughness.

Although no account of polishing gourd food bowls was found, it is probable that they were further smoothed and polished. Covers (*po'i*) for food bowls were sometimes provided; these were made by cutting off the bottom of another gourd with enough diameter to fit easily over the rim of the bowl. Only on the island of Ni'ihau were food bowls decorated; these were called *'umeke pāwehe.* Designs were made up of geometric motifs: lines, triangles in single rows or in combination with lozenges or hourglass shapes, and other simple forms. The designs were comparatively small, with the dyed background more extensive.

Two different techniques were used to create the design. In one, the pattern was obtained by cutting through the outer skin of the gourd and scraping away the part between the decorative motifs. The gourd was then soaked in an infusion of bark, and then immersed in the black mud of a *lo'i.* When the gourd was removed, it was washed in clear water. The parts from which the skin had been removed had turned black whereas the pattern retained the natural beige or tannish color of the original gourd. In the other method, a dye consisting of a mixture of bruised *'ape* leaves and stalks and mud from a *lo'i* was placed inside the gourd. The resulting colored liquid seeped through the gourd wall that had been scraped.

Gourd bowls turn brittle with age and easily become cracked and broken. Hawaiians prized their old bowls and repaired long cracks and even replaced pieces that broke off or out. This was done by boring paired holes on each side of the crack with a bone awl (*kui iwi*) and then sewing the edges together with fine *olonā* thread. The thread was passed twice through a pair of holes, descending obliquely on the inside of the gourd to the next pair of holes. The cross turns on the outside were drawn tight, which brought the edges of the crack close together. Infrequently, the oblique turns were on the outside. Broken pieces were replaced, paired holes were bored around the periphery of the hole, and the replaced piece and stitching done as with linear cracks. This task of repairing gourd vessels was called *pāhonohono* or simply *pāhono*.

GOURD WATER CONTAINERS *(ipu wai* or *hue wai).* There was a greater variety of gourd water containers in Hawai'i than anywhere else in the Pacific; many of these were developed after the Polynesians arrived here. There were three main form groups: (1) a normal globular shape with variations in the shape and length of the neck; those with moderately long necks were called *hue wai 'ihi,* and those with very long necks, *hue wai 'ihi loa;* (2) an hourglass form, with two globular parts separated by a constriction, called *hue wai pueo* because of its resemblance to an owl; and (3) the cylindrical one called *'olo wai.*

A gourd to be used for a water bottle was prepared as follows: at the upper end of the neck the stalk was cut out to form a small circular hole. In some varieties with crooked necks the hole was cut in the side of the neck. Water was poured into the gourd and left there to rot the flesh. After the rotted flesh and seeds were shaken out, small pebbles were added with some water. With continuous shaking the pebbles smoothed down the inner surface of the rind. After this cleaning, a stopper *(pani 'ōmole)* was fitted for the opening; the majority of these were conical shells but they were sometimes made of wood whittled to a conical shape. Decoration of and cord supports for *hue wai* are described in the next section.

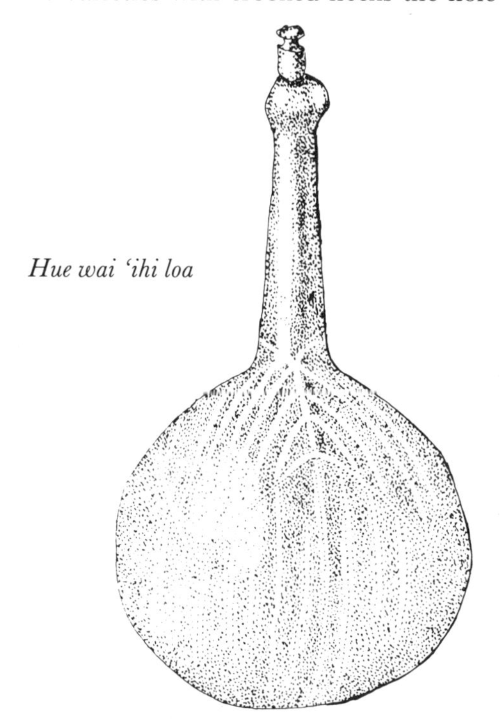
Hue wai 'ihi loa

The globular form with the conical neck was the *hue wai* most commonly used. Those called *hue wai 'ihi loa* were especially handsome with their long tubular necks, more than half the total height of the gourd. In the shorter-necked forms the necks were not much more than an inch high, and in the medium-necked gourds, several inches high. The upper part of the hourglass gourd was usually smaller than the lower part.

Most water gourds were decorated in the manner already described for food bowls.

Suspensory Loops or Bails for *Hue wai (hāwele)*

Because water often had to be brought from some distance to dwellings, it was advantageous to provide some kind of support for hanging the water containers on carrying poles and for suspension from special supports erected inside or outside the house.

Suspensory loops or bails varied from simple loops attached through holes near the rim or tied around the rim if the gourd had an enlarged stalk end, or tied around the middle constriction of the hourglass gourds to more elaborate ones where a series of loops connected a top round below the neck to a double round at the bottom. This was accomplished with one continuous cord; by interlacing the connecting loops no knots needed to be tied. This technique developed by Hawaiian craftsmen of using a continuous cord without knotting is unique in Polynesia. The end of the cord was fashioned into a bail, a loop by which the gourd was hung. The cordage was made from sennit either twisted or braided.

Except for drinking *'awa* from coconut-shell cups, Hawaiians never touched their lips to the rim of a drinking gourd; instead, they held such a gourd elevated above their tilted faces and poured from there into their open mouths.

Wooden Bowls *('umeke lā'au)*

Woods favored for bowls, dishes, and platters were *kou, milo,* and *kamani; koa* was never used for eating receptacles because of a resin that could not be removed, which gives a "bad taste" to food.

Wood to be used for bowls was cut into blocks that were sunk into a pool to soak for a couple of months. If a dark color was desired, the wood was placed in *lo'i* mud. The chunk was then roughly shaped, first on the outside with an adze and then more or less deeply hollowed out depending on its use. This latter operation was accomplished through a combination of tools and methods: after digging out the interior with a stone adze and stone chisel, fire was used to remove the remainder of the internal wood. The sides of the bowl were reduced in thickness, and the entire surface, inside and out, smoothed with successive rubbings using rough lava stones *('a'ā),* coral *('āko'ako'a)* of decreasing coarseness, and pumice *('ana)* or a polishing stone known as *'ō'io;* finally, pieces of stingray or shark skin were used to complete the smoothing process. Sometimes the above processes were finished by rubbing the inside of the bowl with the bract of the male inflorescence of breadfruit.

The final process of polishing the outer surface was accomplished with green bamboo leaves and a few drops of *kukui*-nut oil, or the senescent leaves of breadfruit. Sometimes raw *kukui*-nut kernels, dried in the sun, finely pounded, and wrapped in a piece of tapa, were used in a final rubbing. The latter was also used to polish the inside of the bowl; oiling inside and out made it waterproof and prolonged the life of the bowl. Light-colored bowls from young trees were darkened by using oil from *kukui*-nut kernels that had been scorched or burnt black.

The bottom of bowls was left extra thick so that the bottom weight kept the bowl upright. If needed, a depression was scraped in the ground to fit the rounded bottom of a bowl.

There were two general shapes of bowls of various sizes: broad, low bowls called *'umeke pākākā,* and deep bowls called *'umeke kūmau* or *kūmauna.* The first, although used primarily for *poi,* was also used to serve other foods such as pork, dog, and large fish. The latter was used for storing taro and sweet-potato *poi.* These bowls were provided with low, rounded covers *(po'i)* of wood or gourds to prevent the hardening of the surface of the *poi;* these wooden covers were made in the same manner as the bowls themselves. Medium-sized, deep bowls called *'umeke kū'oho* were used for *poi* to be used soon; they were a common vessel from which the family ate. Also used to serve *poi* to a family group was another medium-sized deep bowl called *'umeke puahala* because its shape resembled a "key" from the *hala* fruit. *'Umeke puaniki* was a general term applied to small bowls that had the same form as larger bowls and were made for individual servings of *poi,* a practice subsequent to the use of the communal *poi* bowl.

Although the majority of bowls had rounded bottoms, some were flat. Also, there was almost always an increasing curve from the bottom to a level at its greatest diameter and then a decreasing curve toward the rim. The greatest diameter of the low, wide bowls was almost always near the rim; in the

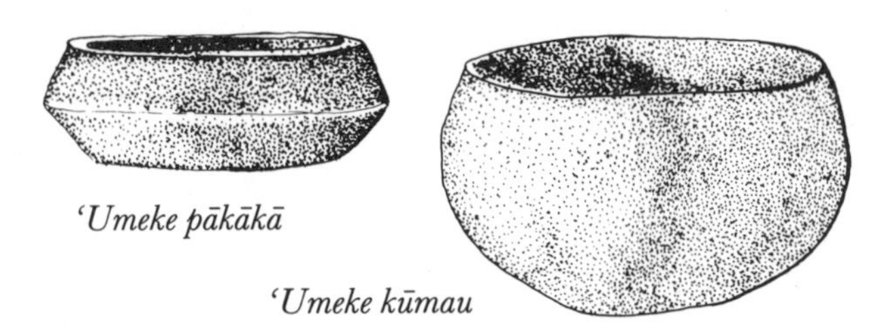

'Umeke pākākā

'Umeke kūmau

Kōkō

deep bowls, it was almost always near the bottom of the bowl.

Variations from the usual continuous smooth outer surface were those with horizontal flat panels, called *'umeke kepakepa,* which were used for *poi,* and those with vertical panels with vertical edges between, called *'umeke 'ōpaka,* usually of small size and reserved to serve *poi* to chiefs.

Special bowls, called *'umeke māna 'ai,* were made for favorite children such as the firstborn *(hiapo)* for use after weaning. These were *kapu* to everyone except the child; no one else was allowed to use these bowls. Bowls of this kind were sometimes made from a *kou* tree that had been planted over the afterbirths *('iewe)* of grandparents.

Heavy bowls to be moved or hung were placed in carrying nets *(kōkō),* which in turn were suspended from carrying poles *('auamo).*

The fine work of Hawaiian craftsmen was displayed in the beautiful carved bowls of various shapes, sizes, and uses, some supported by carved human figures. Meat dishes and platters *(pā)* were of various shapes: elongated, circular, raised on runners, and supported on carved human figures. The main distinction between meat dishes and platters and bowls is that *pā* are shallow. Circular dishes and plates *(pā poepoe)* were also made. The platters with runners were not common; the platter portion was elliptical, with the hollowed-out runners supporting the bottom of the platter lengthwise carved from the same block of wood as the platter proper. These platters were not as large as those without runners. The carved platters resembled the carved bowls except that they were shallow.

Finger bowls *(ipu holoi lima)* of wood were originally circular and elliptical; later they were made in various shapes. A characteristic of many of the older bowls was an internal flange, a convenient edge against which any extra fat or *poi* could be scraped off the fingers before washing them in the water in the bowl. A simple external handle was also sometimes present; this frequently had a hole for a sus-

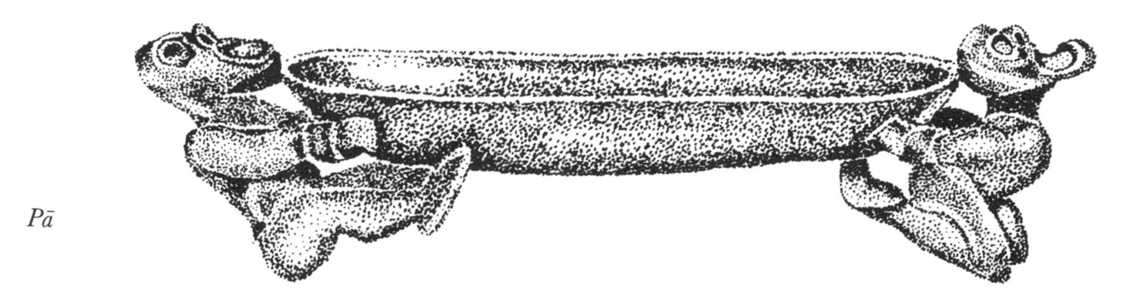

Pā

Ipu holoi lima

pensory loop. The flange was a ridge left when the interior of the bowl was hollowed out.

Scrap bowls *(ipu 'aina)* made of wood were for chiefs to deposit remaining fishbones and scraps of food. It was important that such remainders be protected for fear that a sorcerer *(kahuna 'anā'anā)* might use them to pray the chief to death. These bowls varied from plain to elaborately ornamented and carved ones; all were low, that is shallow, as befitted their use by a high chief. The plain bowls were usually

Ipu 'aina

provided with a short knob that projected from the upper half, some distance below the rim. There was a hole in the knob to accommodate a suspensor loop cord by which the bowl could be carried when empty or hung up between meals. One type of scrap bowl was ornamented with embedded teeth of slain enemies. The feet of the bowl consisted of human figures, carved in one piece with the bowl, in rather grotesque positions.

Repairing *(pāhono)* of cracks and breaks in wooden bowls was accomplished in various ways. In the *pewa* method, a piece of wood of the same kind as the bowl, with an angular shape like an hourglass, now sometimes called a "butterfly" patch, was inserted into a similar hole made in the area of the break but a little smaller so that when the patch was tapped in with a mallet it fit tightly. In the *huini* or *kui lā'au* method, a wooden peg of hard wood, such as *kauila,* was tapped into a hole that had been bored with a bone awl. The *poho* was a wooden patch to fit a hollow part that had rotted or broken out; the rotted part was trimmed to form an even edge and the patch shaped to fit.

Coconut Bowls, Cups, and Spoons

Coconut shells were made into small individual *poi* bowls, *'awa* drinking cups, and spoons. The shell was cut to appropriate shape and size, cleaned of its "meat," and the remaining husk fibers removed by rubbing the outside with increasingly fine coral, next with sandstone, and finally with breadfruit and bamboo leaves. The outside of coconut bowls was given a high polish by rubbing with *kukui* or *kamani* oil. Hawaiians used their fingers to eat most of their food, but for such soft foods as sweet-potato poi and the *piele* desserts they needed spoons, which were cut from coconut shells.

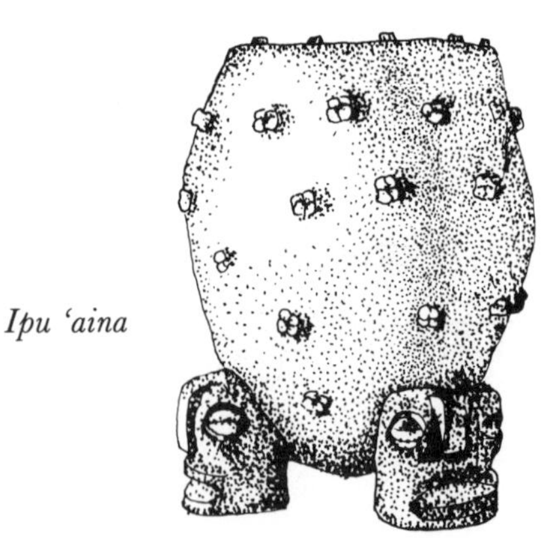

Ipu 'aina

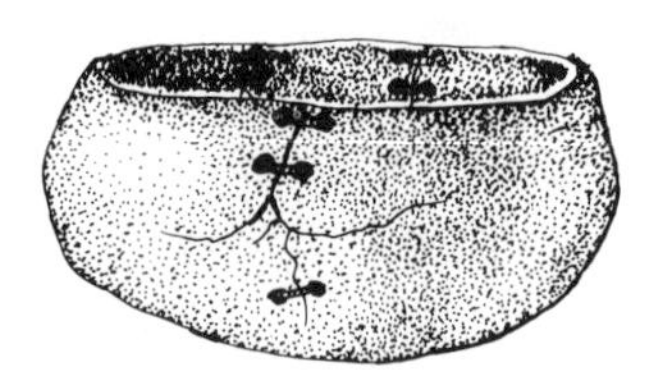

Pāhono

FARMING TOOLS *(nā pono hana na ka mahi ʻai)*

Native wild trees, found here by the Polynesians on their arrival, furnished the hard tough woods used to make farming tools. These trees were *alaheʻe, ʻūlei, kauila,* and *uhiuhi.* Artisans whose specific task it was to make these tools selected the wood and fashioned them with stone adzes and chisels.

The *ʻōʻō,* with its many uses, was the chief farming tool. There were two types of this digging stick: one with a flat point and the other with a flat blade. The thicker end of the stick was cut away at a long slant to meet the opposite side in a flat convex point. In some, the slanting surface was trimmed down on each side of the middle line, leaving a median edge; in others, this edge was missing. The other stick was shorter, with the blade proper constituting half the total length, with slight shoulders and sharp side edges.

The breadfruit picker *(lou)* was a long pole with a short stick lashed obliquely across it at the thinner end to form two acute angles with the pole. Usually a child climbed the trees and "hooked" the fruits down to someone on the ground below who caught them.

A *pālau* was the cutter used to remove the upper portion of the taro corm with the leaf bases *(huli).* There were two types of *pālau:* one consisted of two blades separated by a middle portion for gripping; it had two convex cutting edges. The other type, also double-bladed and with a middle grip, was somewhat longer and had wider blades. In this *pālau* the ends of the blades were rounded rather than pointed as in the other type.

ʻAuamo were poles notched at each end to support balanced loads such as produce from the fields. Such a carrying pole rested on one shoulder. The Hawaiian carrying pole differed from those in other parts of Polynesia in that the sides of the pole were trimmed gradually from the middle. The upper side was trimmed in a similar manner but stopped before reaching the end, thus forming a vertical flange on the upper side at each end.

In the simplest of pole endings, probably the earliest form, the flange formed a terminal knob, without any notch. A common treatment was to make

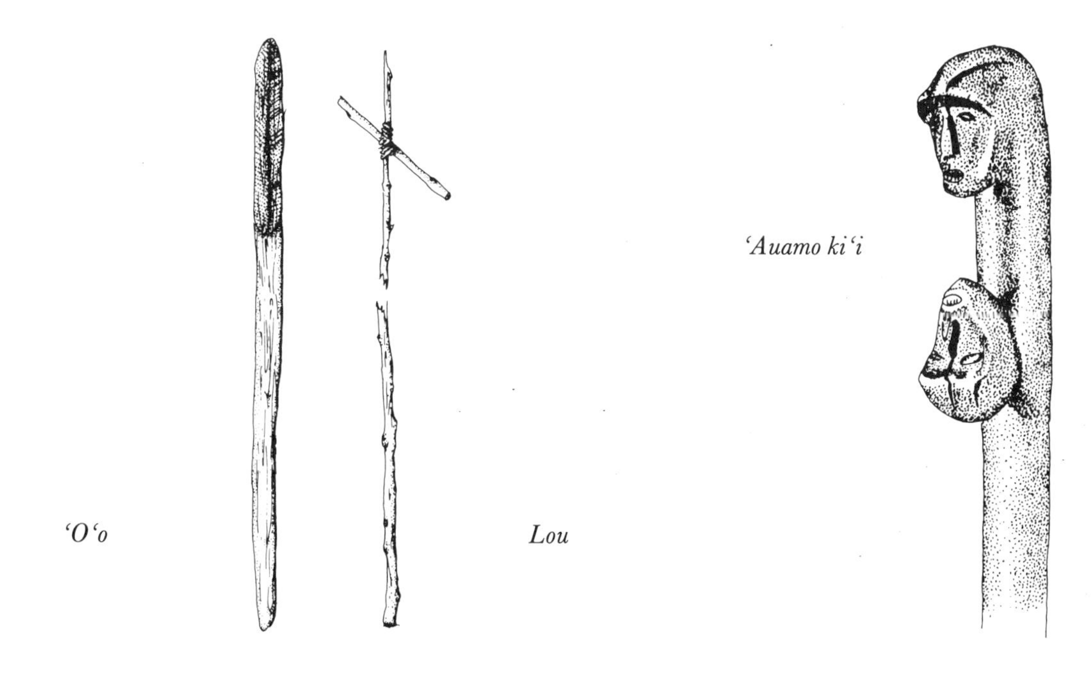

ʻOʻo

Lou

ʻAuamo kiʻi

the flange long enough to permit the cutting of one or more notches through the flange down to the level of the upper surface of the pole. In the households of chiefs of high rank the end knobs were carved into human heads with faces turned upward; these were called *'auamo ki'i.*

The long type of *'ō'ō* was sometimes used for *'auamo.*

SOURCES

Abbot (1982); Bishop (1940, 9–17); Bryan (1938); Buck (1957, 1–2, 5–73); Charlot (1982); Chung (1925); Ellis (1969, 193–194, 215–216, 372–375); Emory (1965a); Fornander (1918–1919, 120–125, 160–169); Handy (1940, 6–196); Handy and Handy (1972, 19–28, 71–188, 198–199); Handy and Pukui (1972, 23–26); Judd (1965); Kamakau (1976, 23–44, 116–117); Kirch (1985, 184–189, 215–236); Knipe (1982); Lawrence (1912); MacCaughey and Emerson (1913); Malo (1951, 22–23, 27–30, 42–44, 51–52, 78–79, 122–123, 204–208); Merlin (1982); Miller (1957); Mitchell (1972, 86–107); Pope (1926); Riley (1982); Whitney, Bowers, and Takahashi (1939); Wichman (1965); Wise (1965b).

2
FIBER CRAFT

It seems best to describe the making of cordage, the twining of baskets, and the plaiting of mats and other articles in a single chapter since the products of these operations appear in several chapters. The making of tapa, however, is primarily associated with clothing and is included in the chapter on wearing apparel.

CORDAGE *(kaula)*

Without metals to make nails and other fastening materials, the ancient Hawaiians used cordage; there were, of course, other uses for cordage. They brought with them, or found in Hawai'i, plants whose fibers were ideal for making excellent cordage. These included the fibers from the husk of the coconut and the bast fibers from *hau* and *olonā.* Of less importance were fibers from the bases of banana leaves, hibiscus, pandanus, ti, *'uki'uki,* and various grasses and sedges.

The fibers from the coconut were pounded free from the husk, the fibers stripped and cleaned of adhering soft tissue between two fingernails, and the clean fibers *('a'a)* spun on the thigh *('ūhā)* by men. The thigh served as an excellent surface for spinning *(milo);* it was a broad, hairless, "oiled" surface since Hawaiians rubbed coconut oil on their bodies as a regular practice. Sometimes the fibers were soaked in salt water before spinning, or placed in the mud of *lo'i* to obtain black fibers. Cordage from coconut fibers is coarse, with a "rough" or "hairy" appearance because of the fiber endings projecting from the body of the cord, but it is very strong and was therefore used when strength and not aesthetics was the main factor.

For cordage made from the bast of *hau,* the bark was stripped from long wands or branches an inch or more in diameter, although older stems or branches were avoided as having fibers too brittle to be useful. For one type of cordage, both the outer and inner barks were left intact and used to make a very strong rope. For this, the entire bark was pulled off in fairly narrow strips; soaked in water to soften the fibers and rid them of a slimy sap; semidried; and then twisted into cords or rope of various diameters. Sometimes the freshly torn strips were twisted into cordage and used in this state. The total-bark *hau* ropes were used to haul the *koa* logs for canoes from the mountains to the sea.

A finer grade of *hau* cordage was obtained from the inner bark or bast. For this type of cordage the outer/inner bark complex was removed in wide strips, the outer bark scraped or pulled off, and the inner bark soaked for a week or more. After this soaking the bast can be separated into five or six separate layers; these are white with a slight sheen and, when held against the light, show a lovely network or lace pattern.

The finest cordage made by the ancient Hawaiians—in fact, the finest cordage made in the Pacific basin—was made from *olonā. Olonā* was cultivated in patches of two or three acres primarily in wet, upland areas. Young shoots or layered cuttings were used for planting material; the latter were obtained by bending down a branch and covering the portion touching the ground with soil so that roots emerged from it. The rooted section, with its terminal leaves, was severed and this became a rooted cutting. Planting was close to prevent side branches from growing. *Olonā* patches were kept free of weeds, especially fom creeping vines, which were abundant in surrounding areas; these would otherwise have choked the *olonā* plants. The stalks were ready for harvest at the end of a year or eighteen months.

Olonā processing sheds were built near running water. The bark was carefully stripped from six-foot-long straight stalks and hung to drain in the shed. The strips were then laid in running water for a day or two only. The strip was then placed lengthwise on a narrow board called *papa olonā* or *papa ʻololī.* This board was shaped to be tapered at one end; this end was driven into the ground to hold it firm. The other, upper end rested on a block of wood to give it a slant. The *olonā* strip was scraped on this board with a tool *(uhi)* made from the backbone of a turtle or a segment of its shell. The bone or shell was prepared by beveling the sides of one end and rubbing them down with a piece of hard coral. The scraper was held in the right hand, and the left hand held the strip of bark against the board. The bark was flattened as the scraper was moved along. This action was continued until the whole strip had been scraped. After the removal of the outer bark, scraping of the bast itself was continued; the resulting product consisted of a mass of fine white fibers. These fibers were then dried in the sun. The fibers were spun before being twisted or braided.

Twisted cordage from spun coconut and *olonā* fibers was made as follows: a long strand was looped around a big toe for anchorage and then each "arm" of the loop twisted separately, either in the opposite or the same direction; twisting in opposite directions minimized unraveling. Twisting was continued until the tension in the two parts of the strand was great enough for one to twist or wind around the other. Two factors determined the diameter of the finished cordage: the size of the individual strands (the number of spun fibers in a strand) and the number of times this process was repeated using the first cordage obtained as the strands for heavier cordage.

For braiding, from three up to five strands, or multiples of these numbers, were braided to make cordage of a greater diameter, or braid of various widths.

For *hau*-bast cordage, individual fibers as for coconut and *olonā* cordage were not used but rather strips of the bast layer. The prepared bast was torn into strips of various widths, depending upon the size of the cordage desired: narrow for finer, wider for thicker.

TWINING *(ʻoai* or *ʻowai)*

Twined baskets of old Hawaiʻi were the finest in Polynesia. The material generally used was the aerial root of *ʻieʻie,* a native relative of pandanus.

All Hawaiian twined baskets *(ʻie)* had circular bottoms. Twining started at the center, worked outward to form the bottom surface and the upward bend, proceeded upward to make the sides, and ended at the upper rim. The elements to accomplish this process were the warps, straight pieces of the root that radiated from the center of the circular bottom and ran vertically or at a slight slant on the sides; and the wefts, which consisted of two or more plies, pieces of root, that were twined around the warps as they crossed at right angles, in rounds, to give structure and shape to the basket.

The Hawaiians used both check twining, the simplest form, where a two-ply weft was twisted around single warps; and twilled twining, where each ply was carried over more than one warp in the same move. There were variations in both methods; these made for pleasing changes in pattern.

Increasing warps to increase the diameter and adding weft plies as those worked with came to an end follows the rules of all twined baskets. Decorative effects were introduced in various ways, such as adding a single or more rows of twill in a basket twined in check.

Because most twined baskets were fitted with covers, loops of cords were incorporated in the twining as the basket approached its full height or length; these loops formed the foundation of a net that would secure the cover. The most common manner of incorporating such loops was to run an *olonā* cord, in a series of loops, around two warps with about six warps between. A variation was to loop the cord around one warp, tie a knot and proceed as above. A less-common method was to lay the ends of separate loops vertically against warps, fixing them with subsequent rounds.

Finishing the rim entailed preventing the top, last round from slipping over the top of the warps and, in the better baskets, giving a neat finish to the rim. To accomplish these objectives the Hawaiians used variations of two main forms, the braid and the wrapped hoop. In the first, the composite warps were bent forward and outward to cross the top ply of a three-ply weft as it passed inward to an interwarp space; the bent warp was then crossed in turn to the other two plies of the weft. The wrapped hoop finish was formed by placing a double hoop of unsplit *ʻieʻie* root on the upper edge of the rim and then wrapping a length of split root in close turns around the hoop to cover it completely.

The variety of baskets can be divided into four types: type I were baskets with widths markedly greater than the height; two variations were (a) squat and (b) round. In the former, the greatest diameter was near the lower end and the basket narrowed upward toward the rim, with the height slightly more than half its greatest diameter.

Type II baskets were elongate, with the height greater than the maximum width. They had both globular and flattish covers, and were provided with net meshes attached to the cord loops already mentioned to keep covers in position. These, the largest group of basket forms, were of three kinds: with contained gourds, with contained wooden bowls, and those without containers.

For the gourd baskets, twining was close around a selected gourd of medium height, with its stem cut off to provide an opening. The bottom of another gourd formed a cover over the opening of the main gourd; it was globular and was also enclosed in twined basketry.

Wooden bowl baskets were of two kinds: those with covers and those without. As with the gourd baskets, the twining was done close to the outer surface of the enclosed container. The cover was also of wood and was covered with twining.

Although some baskets were made without containers, some of these probably lost their gourd container. There were baskets that were definitely not made to cover gourds; they do, however, follow the shape of gourds. These differ from the gourd baskets in possessing covers that have flattened instead of

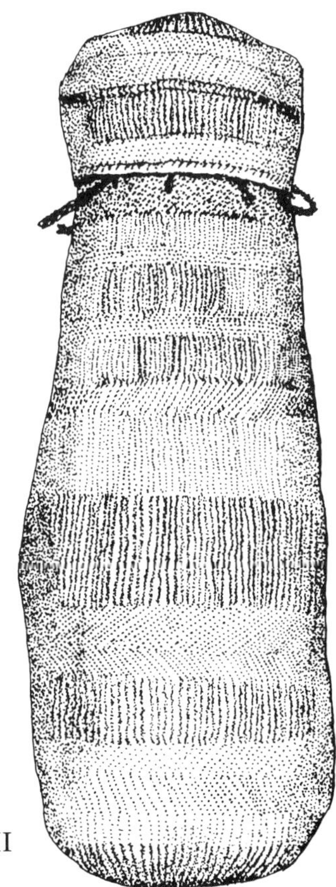

ʻIe, type II

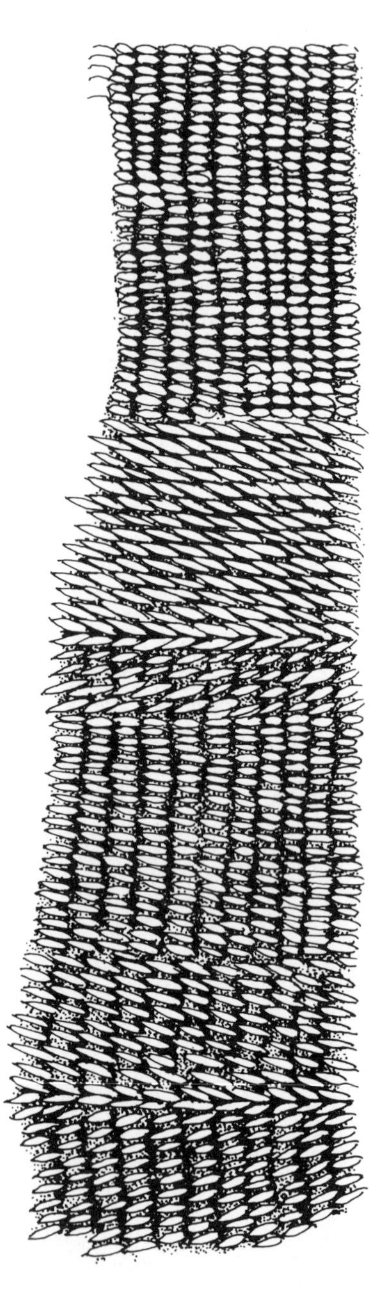

'Ie, detail

'Ie, type III

domed tops, they are shorter, and have larger openings than the other two kinds.

Type III baskets were more coarsely made than those of types I and II, with the weft rounds farther apart. The bottoms were flatter and their shapes did not follow the contour lines of gourds.

Type IV baskets included small containers, more or less bottle-shaped. The material for this type of basket was dyed brown.

The large baskets of type I were used to store tapa and other valuables. The baskets of type II, with containers, were used to store *poi;* the container held this semiliquid food and the basket protected the gourd, which became brittle over time. The coarser baskets of type III were used as containers for fish, and the small baskets of type IV served as containers for small fish and shrimps.

Decorative effects were introduced in baskets by changing twining techniques, but even more distinctive results were obtained by using weft plies that had been dyed black. It was the better-made baskets of types I and II that were usually decorated in this manner. As already stated, some of these were twined entirely with brown material, thus making an attractive background for the black decorative designs. These might be alternate columns of brown and black or a narrow band of the body color in plain check between rows of twill in black. The weft plies were dyed black by placing them in the mud of *lo'i.*

PLAITING *(ulana* or *nana)*

Parts of several plants were used for plaiting: leaves *(lau hala)* and the bracts of male inflorescences *(hīnano)* from the screw pine, *hala;* the stem of the great bulrush, *'aka'akai;* the stem of a fine sedge, *makaloa;* and the leaves of fan palms, *loulu.*

The main use of *lau hala* was for making mats. The ancient Hawaiians used leaves at different stages of development, in contrast to today when only dry leaves are used: from older, dried, tan-colored, tough leaves for ordinary mats to younger green leaves and white unopened leaves in the center of leaf clusters. These last were used for the finest *lau hala* mats.

The green leaves were drawn over a fire to wilt them, lighten their color, and make them more pliable. The dry leaves were prepared as follows: the base *(po'o lau)* and the tip *(welelau)* were cut off, the former where the leaf narrows and the latter several inches from the end; the marginal spines removed *(kīhae)* by inserting a thumbnail or perhaps a splinter of bamboo inside the spiny margin, or simply ripping; and the spines on the back of the midrib scraped off with a shell. The leaves were then wiped with a piece of moist tapa or the cut-off base of the leaf to remove any dirt. The cleaned leaves were rolled over the hand to flatten them, a method still practiced today. The leaf was rolled twice, once from base to tip and then from tip to base.

After this smoothing and softening process, several leaves were laid one upon the other, the bases held in the left hand, further flattened, and then rolled *('ōka'a)* into a small roll and tied. Later these leaves were sorted, made into large rolls, tied, and put away for later use; such large rolls were called *kūka'a.* Another method of preparation was to hang flattened leaves in a protected place for several weeks, give them a second softening, and then make them into the large rolls. The best time for this preparation of these older leaves was early in the morning, or in the evening, or on rainy days when the air was moist.

Bleaching of mature leaves was accomplished by exposing them to the sun during the day; they were put under shelter at night or when it rained. This process was repeated for days until the desired degree of bleaching was obtained.

When ready to plait, the prepared leaves were further softened by piling the leaves one upon the other, with tips and bases alternating, on a stone and beating the pile with an old wooden tapa beater. A supply of strips for plaiting was prepared by splitting the prepared leaves *(kīhae lau hala)* with a bamboo, wooden, or bone splinter into the required width *(koana);* a working supply of these were prepared before any plaiting was commenced.

The chief plaited article made from *lau hala* was the floor mat. There were different types of these depending upon the leaves used and the width of the strips; mats made with wide strips, an inch wide, were called *moena lau,* and those made of strips only a fraction of an inch wide were called *moena makahune.*

Mats were plaited either with single or double wefts; floor mats were usually plaited with double wefts to withstand the wear of walking over them. Two different patterns or methods of plaiting were used: checker (board) and twill (diagonal), with the latter preferred for mats.

The beginning of plaiting; the formation of the left lower corner; the defining of the side and bottom edges; the addition of wefts; the turning of the right lower corner; the defining of the right side edge; the building up by short sections; the joining of wide sections; the formation of the top left corner; the top finishing edge with the disposal of weft ends; and the completion of the work at the right top corner—to describe all these processes in making a mat would involve pages of description and lies outside the scope of this book. Such descriptions appear in several of the references given at the end of this chapter.

Coarse sedge mats called *moena ʻakaʻakai* or *naku* were made from the native great bulrush (*ʻakaʻakai*). The dried stems were flattened and plaited into mats for the lower layers of the large Hawaiian couch (*hikieʻe*) and for temporary mats for various uses, temporary because the material is not durable.

Although *makaloa* grew on all the major islands, it was chiefly on the island of Niʻihau that mats were made from this sedge; hence the name frequently used, Niʻihau mats. The Hawaiian names for the mats are *moena makaloa* or *moena niʻihau,* and *moena pāwehe* for those with a colored geometrical pattern on the upper surface. *Makaloa* mats are considered the finest sleeping mats in Polynesia.

The tubular stems of *makaloa* were flattened as they dried, resulting in strands with shiny outer surfaces on both sides. Plaiting of the strands was in the common check pattern although sometimes sections of twill were introduced. The technique of plaiting these mats is somewhat different than that used in plaiting *lau hala.* Again, details for the plaiting of these mats may be found in the references listed.

Makaloa mats were not only famous for their fine plait but because of their beautiful decoration. This consisted of the introduction of colored geometrical motifs on the upper surface by overlaying colored wefts on the foundation wefts and plaiting them as double wefts for the areas covered by the design. The colored wefts were laid on both the left and right crossing wefts to obtain a solid motif in one color; they were cut off at the far edge of the last crossing weft. The colored motifs are visible only on the upper surface of the mat since the colored wefts were applied on the upper surface of the foundation wefts. The colored wefts were obtained from the sheath at the base of the sedge stems; these are naturally red when fresh but in time turn brown.

The generic term for the colored motifs in *makaloa* mats was *pāwehe,* thus the name *moena pāwehe* as an alternate name for these mats. The geometrical designs consisted of lines, usually zigzag (*keʻekeʻe* pattern); triangles, such as one with opposing triangles with their apexes touching to give the appearance of hourglasses (*papaʻula* pattern); a horizontal row of triangles each with its base vertical and its apex touching the base of the next triangle in the row (*puahala* pattern); and two vertical rows of triangles with the bases below and the apices touching the bases above (*nēnē* pattern). Another motif was produced with lozenges, such as red bands that were internally enhanced with a continuous row of white lozenges (*humuniki* pattern), and contiuous rows of lozenges with all angles touching (*papa kōnane* pattern).

Moena

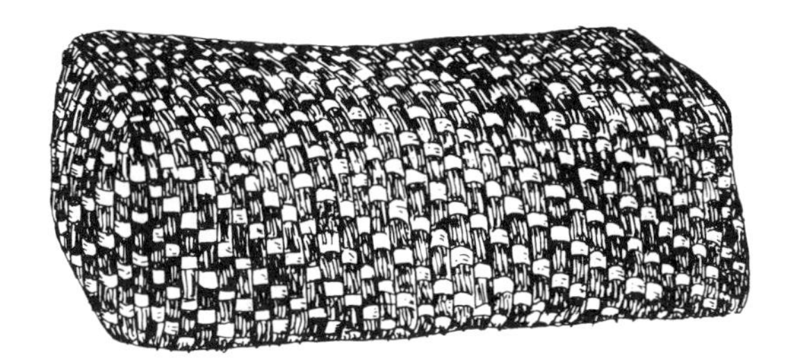

Uluna plaited from *lau hala*

As already noted, the leaves of the native fan palms were also plaited. The blade tissue between the radiating veins of very young leaves was torn into strips and plaited into fans and baskets. These were both made by the same technique used in making fans and baskets from *lau hala,* described below.

Besides mats plaited from pandanus leaves, other articles such as pillows *(uluna);* baskets *(hīnaʻi)* of various sorts, and fans *(peʻahi)* were also fashioned from this material.

Pillows were either rectangular or cubical in shape, varying in dimensions. They were usually plaited in check, although some pillows were quite elaborate, plaited in twill with ornamental overlays of dark material, such as banana skin. They were stuffed with pandanus leaves that had been folded into shorter lengths to the approximate width and depth of the pillow. Separate leaves were folded to accommodate the length of the pillow and then the two bundles were bound together with some longitudinal turns of another leaf before being covered by plaiting.

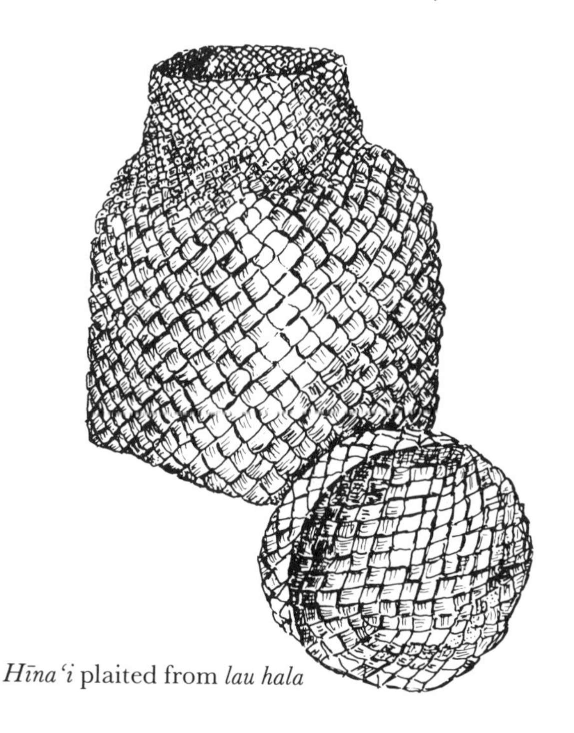

Hīnaʻi plaited from *lau hala*

There were two types of baskets made of plaited *lau hala:* a round one without a cover and a round one with a cover. A satchel form of *lau hala*-plaited basket is of postcontact origin. Details of the techniques involved in plaiting baskets may be found in the references given.

Fans were plaited from both the leaflets of coconut and from *lau hala.* These varied in shape; the simplest was triangular in form. A simple fan was plaited from coconut leaflets in a simple check, with the ends of the wefts gathered together and tied with sennit to make a handle. In another simple fan the handle was formed of a section of unsplit midrib at the tip of a frond; the attached leaflets were first braided over the midrib and then plaited to form the body of the fan. Elaborate fans were made for chiefs; these appear to be more ornamental than useful. They were usually made from the pandanus leaf, closely and neatly plaited in contrast to the simpler ones that had wider wefts and were loosely plaited. Of these finer fans some were arch-shaped, others square. They had spreading handles that were embroidered with human hair and some unidentified brown fiber.

SOURCES

Bishop (1940, 19–22); Buck (1957, 111–140, 148–164, 290, 312–323); Fornander (1919, 626–629, 656–659); Kamakau (1976, 61, 64, 79, 89, 98, 105, 113, 117–118); Kirch (1985, 193, 195); Malo (1951, 22, 49–50, 78–80, 210–211, 213); Stall (1953); Stokes (1899, 27–28, 31); Summers (1988, 24), (1990).

3
FISHING

Polynesians arriving in Hawai'i rejoiced at the quantity and variety of both marine and freshwater fish they found; as with all Polynesians, fish provided the main source of protein in their diet.

Those who fished in old Hawai'i had many skills, even more than farmers and a great many more tools. There was much for these people to learn: the character of the bottom of the sea, both inside and outside the reef; the contour of the reef itself; the clefts and crevices where fish hid not only in the reefs but in the rocks of the shore; *ko'a,* the deep-sea fishing grounds; and much more such as the right kinds of nets and the proper hooks and bait for each fish.

Something should be said about the fish watchman *(kia'i);* he was so skilled that he could tell where a school of fish had gathered through signs unseen by the ordinary person: a darkening of the sea, the slightest ripple beneath the surface, and many other indications.

METHODS AND TOOLS

There were many methods of catching fish: by hand, spearing, noosing, netting, trapping, and with hook and line. The last three especially required many different kinds of equipment.

Groping

Catching by hand or groping *(hāhā)* was the simplest method of catching fish and was used by both men and women around rocks in shallow water. The fisherman or woman dived down in likely places and thrust his or her hands under rocks and in holes, and under coral projections from where fish could not escape. Fish so caught included *hou; 'ōlali,* the juvenile *hou;* and *pakauele,* a general term for those fish easily caught with the hand, such as *manini,* a common reef surgeonfish; *lauhau,* a bright colored small fish; and *pāo'o,* designating different varieties of *'o'opu.* Crayfish were also caught in this manner.

Spears and Spearing, and Noosing

Fish spears *(kao* or *'ō)* were straight sticks made (entirely or their tips) from such woods as *kauila, uhiuhi,* or similar hard woods. They were slender and sharply pointed or barbed at the tip, with the other end, the butt, trimmed to a lesser thickness than the shaft proper. This spear was used primarily for different kinds of rock fish; the experienced fisherman or woman knew the likely spots near large coral rocks or on the steep faces of reefs. It was with this spear that fish were speared inside the reef at night with the aid of torches *(lama lama).* The latter were made from a length of bamboo including an internode, a node, and part of another internode to serve as a handle. The longer internode portion was filled with dried or roasted *kukui*-nut kernels; holes were bored in the walls of this section to admit air necessary for the nuts to burn. Ignited, the nuts gave off a smokey but bright yellow light that attracted fish. This spear was also used to catch porcupinefish *('o'opu hue)* and squid *(mūhe'e)* in shallow water.

A deep-sea gray shark *(niuhi* or *manō kanaka)* was

caught by noosing as follows: on successive days, bait consisting of pounded baked dog or pig liver and *'awa* wrapped in ti leaves was carried out to sea in canoes to deep water where bundles were dropped. Each day it was dropped nearer to the canoe, until one day when the shark had become satiated with liver and stupified from *'awa* a noose of sennit could easily be slipped over the shark's head and it was towed to shallow water where it was killed. Catching this shark was the special sport of chiefs; he who slipped the noose over its head acquired good *mana,* great prestige. The eating of the flesh of this shark was *kapu* to women.

Nets and Netting

Nets *('upena)* were made in various sizes and shapes because netting was the most diversified and rewarding method of catching fish. *Olonā* was the most frequently used twine for nets as it is very strong and does not kink. For shark nets, however, the undressed bark of *hau,* a combination of the outer and inner barks, was used. Stems of a native sedge, *'ahu'awa,* were used for the large-meshed turtle nets. Another source of fiber for netting cordage was *hōpue* or *ōpuhe.*

The implements used in making nets were the mesh gauge *(haha kā 'upena),* the netting needle or shuttle *(hi'a kā 'upena),* and the net mender *(kī'o'e).* Gauges were made of wood, turtle shell, bone, and whale ivory. Wooden ones were made either of a piece of bamboo or from the bastard sandalwood *(naio).* They were rectangular, with corners of one or both ends usually rounded off. Some had a hole through one end to accommodate a cord loop.

The netting needles or shuttles were made of bone, whale ivory, or *kauila* and *naio* wood. They consisted of a shaft with a large slit eye at each end, the intermediate portion between eyes having been thinned down to make a rounded shaft. Each eye was formed of two curved arms enclosing a circular or elliptical space. These eye ends were slit to admit the cordage for winding between the eyes. Wooden shuttles were usually larger than those made from bone or whale ivory. For making large-meshed shark and turtle nets, no shuttle was used. Instead, the cord was wound over the hand and elbow for several turns as for a hank of wool in knitting; the hank thus formed was doubled and wound with the rest of the cord until a pear-shaped ball was formed. In making the net, the cord was drawn from the inside, through the point of the ball; thus the ball retained its shape until all cordage had been drawn out. The ball was passed through the large meshes that were spaced by drawing down the cord with the left hand to the required mesh size and then making the netting knot with the right hand.

Hi'a kā 'upena

Net menders were straight pieces of wood with a thicker handle cut down to make a sharp shoulder at its junction with the thinner point portion.

To set up the first and additional rows of meshes, the shuttle was filled from a ball of cordage and left connected with it. A separate length of cordage was made into a loop that was stretched between the netter's two big toes. Then, in a combination of passing over and under the toe loop and behind and in front of the mesh gauge, the shuttle brought the cord into position to make a characteristic fisherman's or English knot. The first mesh was completed. The first row of meshes was continued until the number required for the particular net was obtained. The ball cordage was then cut off at the last netting knot and the first row swung around so as to be ready to start the second row, again working from left to right.

To join edges *(pāku'iku'i)* of certain parts of bag nets, five methods were used: (1) a separate cord *('aea)* was run through the meshes of the two sides in turn; (2) using a separate cord, opposite meshes of the two edges were tied together with an overhand knot; (3) a separate cord joined alternate meshes on each side with an overhand knot; (4) a row of meshes was made along an edge and the shuttle run through the mesh loops on the other edge; and (5) the two edges were joined together with netting knots with the shuttle.

GILL AND SEINE NETS. The chief difference between gill and seine nets was length. Both nets were fitted with stout cords or fine rope, called *'alihi,* which were threaded through the marginal meshes of the upper and lower edges. The upper, called the head cord, had floats *(pīkoi)* attached to it and so was called *'alihi pīkoi.*

Net floats were made in three different forms: cylindrical, D-shaped, and wedge-shaped. The cylindrical floats were made from a short length of *hau* branch from which the bark had been removed and a hole bored lengthwise through the center. A binding cord of *olonā* was passed through the hole and tied at each end to the head cord of the net. D-shaped floats were made by splitting cylindrical sections of *hau* lengthwise; thus one side was flat and the other rounded. These floats were usually longer than the cylindrical ones. A hole was bored transversely toward each end and a binding cord was passed through one end hole, around the head cord in a number of turns, and then around the binding between the float and rope. The binding cord was then carried along to the other hole and a similar tie made. Wedge-shaped floats were made by cutting wedge-shaped pieces from circular sections of *hau* wood; the base of a wedge was formed from the outside of the branch. The two sides were cut down at a slant to meet a blunt edge. The ends of the base were rounded off and a hole bored at each end near the blunt edge of the wedge. This float was fastened in the same manner as D-shaped floats.

The lower cord, called the foot cord, was fitted with stone sinkers *(pōhaku)* and was therefore called *'alihi pōhaku.* Different types of sinkers were used on gill and seine nets than on hand nets and on lines.

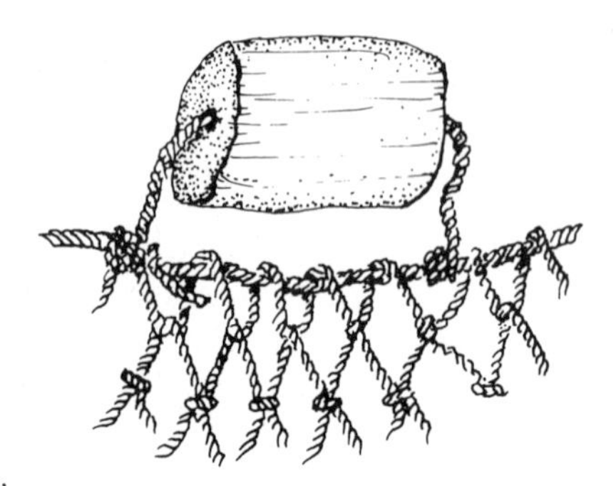

Pīkoi

Those used on gill and seine nets were usually made from vesicular basalt, which was easier to shape and groove than the denser basalt and coral. They were usually elliptical, with transverse grooves to hold the binding cord.

The generic name for gill nets was *'upena ku'u;* such nets were distinguished in the way they were used. *'Upena ho'olewalewa,* literally hanging net, was stretched in shallow water at high tide between two points across fish runs with a long sandy opening in coral areas. One or two persons worked the net, moving backward and forward on the seaward side to take out the fishes caught in the meshes. This method was used only at night, or the net was left overnight and the fishes caught by their gills during the night were removed in the morning. *'Upena 'apo'apo,* literally encircling net, was used to encircle an area in which fish were seen. Fishermen entered the circle and beat the water, usually with a coconut frond, to drive fish into the net, where they were caught by their gills.

As already noted, seine nets were similar to gill nets but were much longer. They also had head ropes with floats and foot sinkers. A very large seine net, to catch *'ō'io,* lady- or bonefish, was used in deep water near the reef. Fishermen went out in two canoes; the lookout stood in the bow of one of the canoes watching for a school of fish and when he sighted one the canoes followed slowly and quietly until the fish moved into a sandy-bottomed fishing ground called *ku'una.* One canoe cautiously ap-

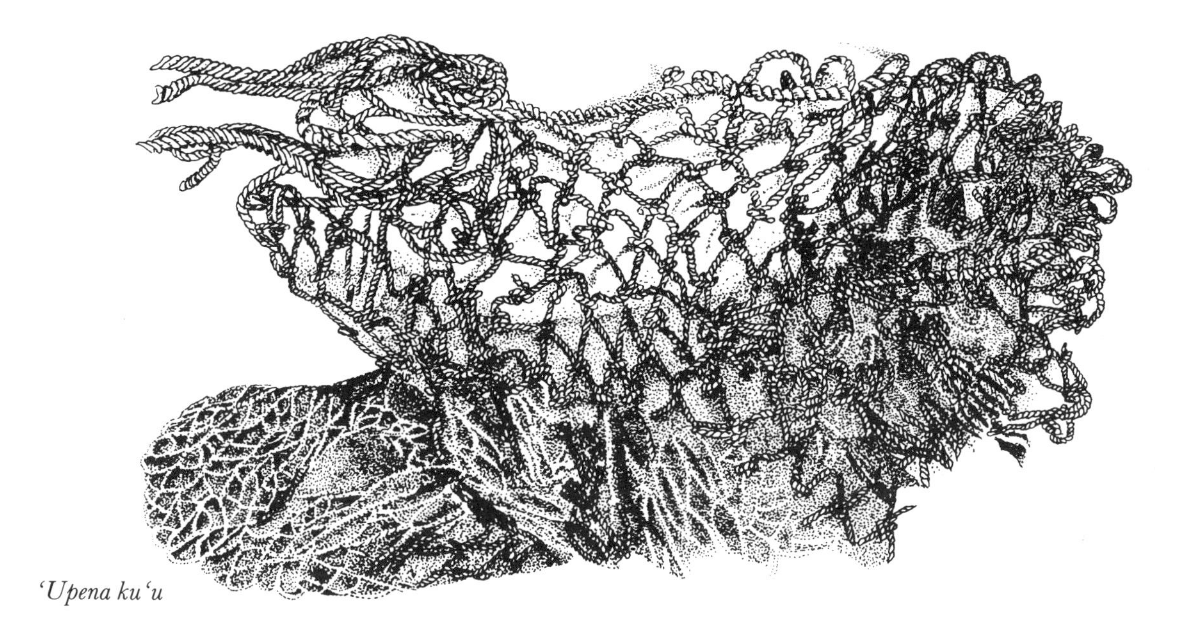

'Upena ku'u

proached the place where their net was to be dropped. Another canoe with a similar net made a wide circuit until it was opposite the first canoe with the school of fish between. One end of each net was dropped simultaneously, and the two canoes were paddled swiftly in a semicircle to meet the dropped end of the other net, paying out their own net as they paddled. The first canoe to reach the end of the other net kept on paddling on the outer side of the circle, still drawing its net after it to decrease the size of the circle. Two or three circles of netting were thus made around the school of fish so that if they escaped over the top of an inner circle an outer circle would stop them. When all was ready, the head fisherman gave a signal, upon which all the fishermen jumped overboard into the circle and beat the water with their hands or paddles to drive the fish into the net(s), where they were caught in the meshes.

There were other types of seine nets: *'upena kākā* was used by two men to catch mullet or *'ō'io;* a very fine-meshed net was used to catch small mullet and *awa* (milkfish), both to stock fish ponds; and another to catch larger mullets, *awa,* and *pa'ū'ū ulua,* the second growth stage of *ulua.*

HAND NETS. Hand nets were small bag nets supported by a frame made from a branch of *'ūlei* or some other tree, and used by a single person as a scoop net. There were two forms: one with an oval frame, ending in a point at one end; this net was called an *'upena,* further distinguished by an adjective designating the fish caught with it; the other, with a frame consisting of two straight sticks, was called *'upena uluulu* or *'upena ahuulu.*

The general structure of the oval-frame scoop net was a length of branch bent into a distal curve, with the ends brought together to make the shape of the loop. Usually a single length of branch was used, but if necessary, as for large frames, extra pieces were spliced on the sides. A short crossbar, lashed to the two sides of the frame near their junction at the pointed end, fixed the shape of the frame.

A smaller net was fitted with netting made of sennit. This was fastened with a continuous cord to the frame, at intervals, with double half hitches through the marginal meshes. Along the front of the crossbar the net was fastened by passing a double cord through the marginal meshes, which was then tied to the frame on each side. This type of net was used for scooping up fishes that had been surrounded by seine or gill nets.

In a larger net the net portion consisted of up to fifteen pieces of fine-meshed netting of different sizes joined to form a bag, with its greatest depth near the handle. The upper strip of netting was made to fit the circumference of the frame and was fastened to this by running a cord through the upper marginal meshes and fixing it to the frame with half hitches at intervals. The netting for this net was made of *olonā* cordage. Such a large net was used to catch *pāoʻo* (better known as *ʻoʻopu*) and *ʻiao* or *ʻiʻiao* (silversides).

In some scoop nets the frame was made of twisted vines, probably *ʻāwikiwiki,* made in the same manner as with branches in the other two types described. There was no crossbar in this type of scoop net. Such nets were used by women to catch small rock *pāoʻo* and *ʻōpae,* freshwater shrimps, usually at night.

In the two-handled scoop nets two parallel sticks were used instead of a hoop for the frame. The netting was usually a rectangular piece cut from a larger net made from cordage from either dyed *wauke* or *olonā* fibers. The piece of netting was doubled and the two open sides closed by making overhand knots on alternate meshes of the two edges with a piece of cord. The two sticks were inserted through the marginal meshes of the two upper edges of the netting. At the far end, the two sticks were tied with a cross cord, which passed through a marginal mesh; this left an interspace between the two sticks. At the other end, the sticks extended beyond the netting to form handles. Sizes and netting mesh varied. Such nets were called *ʻupena uluulu* and were used by men who dived down to small caves or holes; here they placed such a net across the opening and, with the help of a leafy branch *(pula)* drove the fish out into the open mouth of the net. The diver-fisherman then brought the two sticks together to close the net opening.

ʻUpena ʻākiʻikiʻi were dip nets of the "orthodox" type; these consisted of a square piece of netting tied at its four corners to the arched ends of two crossed sticks and suspended from a line tied to the crossing of sticks. The net did not form a bag until the weight of the fish brought the ends of the sticks closer together as the net was drawn through the water. Some dip nets such as those used to catch *pahuhu,* a young stage in the growth of *uhu,* were so large that the arched supports were made of spliced sticks, which were lashed firmly together with diangular turns with either sennit or *olonā* cordage. In a typical net a marginal cord was run through the meshes of the four edges of a square net and tied at the four

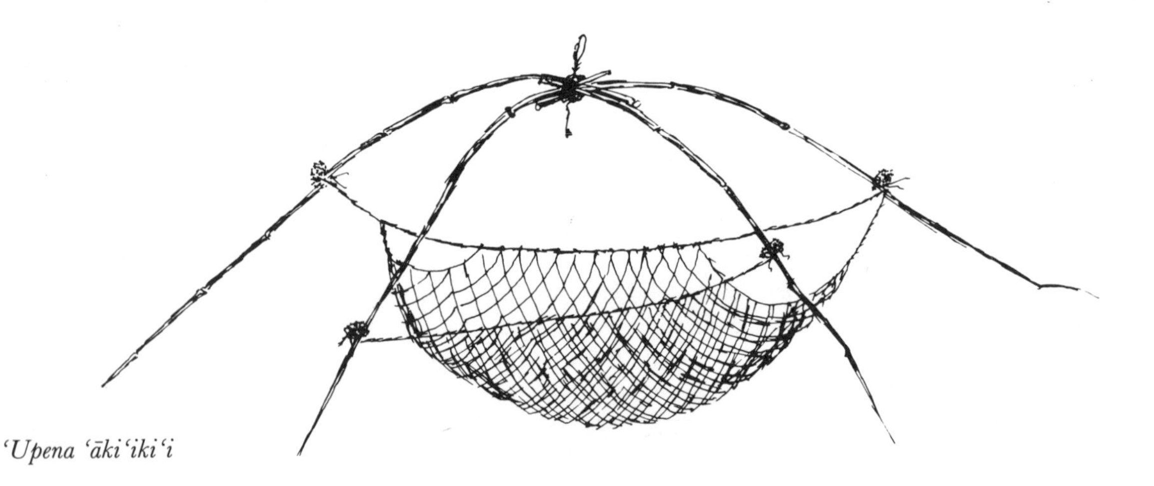

ʻUpena ʻākiʻikiʻi

corners so that a loop of the marginal cord formed at each corner. These four loops were tied to the lower parts of the sticks. The four ties drew the four limbs into taut arches. Stone sinkers were tied to the lower part of the sticks with transverse turns around sinker and stick a few inches from their ends. Variations from this more usual type included those with circular or rectangular pieces of netting, and netting with different meshes.

Larger nets of the same general type were made to catch full-size *uhu;* they were called *'upena uhu* or *kākā uhu.* The *uhu* net was used in water where the bottom could be seen after chewed *kukui*-nut kernels were spat over the surface to smooth it. Fishermen, using live *uhu* caught on a hook as a decoy *(pākali)* attracted other *uhu* through a process called *ho'ohaehae* (teasing). The net was lowered to a short distance from the decoy, which was gently drawn on to the net. When a fish followed the decoy into the net, it was quickly drawn up. The flexible arches bent over as the weight of the fish caused the net to sag into a bag. There was a trick in pulling the net string so that the two arched sticks swung together from the crossed position.

Bag nets. There were several types of bag nets, of various sizes and kinds of enclosures: *'upena nae kuku, 'upena 'ōhua, 'upena papa, 'upena kolo,* and *hano mālolo.*

'Upena nae kuku were made from large rectangular pieces of netting. Although large and small nets were folded differently—the larger ones were folded from below upward to form the bag; the smaller were folded over from right to left to make the bag—the result was the same. There were other ways of folding and variation in sizes and numbers of pieces used to make the final rectangular netting piece.

To form a bag, the bottom or one end was closed by the marginal cord method; thus the bag was open at the top and along one edge. A strong cord, similar to the *'alihi* cords of gill and seine nets, was attached along the free edges. The *'alihi* cord on the left functioned like the foot cord in the gill net with sinkers; the upper one, though joined at right angles to the other, functioned like the head cord with floats. When the net was in use a stick *(kuku)* was used at each end of the foot cord.

These nets were used as follows: the foot cord with its sinkers was stretched to its full length on the sea bottom. The lower ends of the two sticks were stuck through a loop at each end of the foot cord. The sticks, also attached to the adjacent sides of the head cord, were held upright or at a slant by two men while the rest of the head cord with its floats curved behind to form an irregular opening to the shallow bag. A thick bushy rope made from the twisted vine of the beach morning glory *(pōhuehue)* was attached to each of the sticks. The vine ropes were pulled forward in a semicircle by fishermen, sweeping shoals of fish before them. When enough were enclosed, the two ends were brought together to form a circle, the circle was decreased by overlapping the ends, and finally the fish were driven into the net. This net was used primarily to catch mullet fry for stocking fish ponds, or other small fish for bait.

'Upena 'ōhua were nets made much like the nets just described; they were used to catch small fish, *'ōhua,* greatly valued by the Hawaiians. The operation of this net was similar to that in *nae kuku* fishing but was on a grander scale. Long ropes, up to several hundred feet, on which ti leaves were braided by their stems, with blades hanging loose and free, were pulled in opposite directions from a given point to sweep around to form a circle. This circle was reduced by making concentric circles. Men, women, and children, in great numbers, took part in this kind of fishing: keeping ropes down on the sea bottom, driving fish away from ropes by splashing, and disentangling ropes if they got caught on rocks. When the circle had been reduced to a diameter of ten to fifteen feet, one end of the rope was untied and the ends attached to the ends of the *kuku* of the bag net; this formed a guard on either side.

After further reduction of the circle, fish were driven into the net.

'Upena papa were larger than *'upena 'ōhua.* They consisted of a main or funnel part and had an additional cylindrical bag that extended another several feet beyond the narrow end. Although the free end was open, it had a cord attached to it to close it when the net was in use. There were variations in shape and size of these bag nets.

These nets were operated in the following manner: two long ropes, over a thousand feet in length and prepared with ti leaves in the same manner as the ropes in *'ōhua* nets, were placed, along with the *papa* net, in a double canoe and taken out two or three miles from shore. In this canoe also sat the head fisherman. Accompanying his canoe were as many as sixty to eighty single canoes, each with eight to ten fishermen. The two ropes were tied together and men in two canoes, each with a free end of the rope, paddled off in opposite directions, parallel to shore. The double canoe remained stationary at the point where the ropes were joined. The remainder of the canoes divided into two groups, those in one stationing themselves at intervals along one rope, those in the other, along the other rope. Once the rope had been laid out to its full length, the men in the leading canoes pulled the rope with all their strength toward shore with the other canoes also moving forward toward shore. Near shore, in a clear sandy area, the rope was untied and attached to each side of the *papa* net, which had been dropped into the water from the double canoe. By making a great disturbance with their feet, men, women, and children drove fish into the net, which, along with the rope, was drawn to shore. These nets were used to catch various types of flying fish as well as other fishes.

'Upena kolo were immense, deep bag nets made from small-mesh netting, narrow at one end and with a very large flaring opening. Attached on either side of the bag net were long nets called *pepeiao* (ears). Because of the depth of this combination net it could be used in only a few places, such as in Honolulu harbor. In use, it was swept from one side of the harbor to the other, thereby scooping up all kinds of fish. It was designed for this very purpose, to catch large quantities of fish. Necessarily, a great many persons were needed to handle such a net.

Hano mālolo, the net made to catch flying fish, consisted of two parts: a stretch of open netting, narrowing in the middle and narrowest at one end where it was attached to a second part, a tubular device. The latter was a bag with an arch to complete the tube that was open at its free end. A strong cord was attached to the meshes near the outer opening to close it when the net was in use. The open portion of this net was made of cordage of ordinary weight while the bag portion was made of thicker cord.

This net was used as follows: it was taken out for miles, to deep water, in the canoe of the head fisherman, who followed the lookout in his light canoe. These two head canoes were followed in turn by a fleet of as many as twenty to forty canoes, often paddled by women. When the lookout saw a strong ripple on the water, indicating a shoal of flying fish, he pointed to the spot and the other canoes surrounded the area. Next, the head fisherman dropped the net lengthwise, with the bag part kept near the surface with floats. Ropes tied to corners of the lower end of the net were held by men in canoes, one on each side, some distance apart. With the net so spread, the other canoes were paddled toward it, the paddlers splashing water to drive the fish inside. Men in the two canoes pulled vigorously on the ropes to bring up the lower end, and the fish were driven into the bag.

Traps

Fish traps *(hīna'i)* of various sizes, shapes, and materials were made for catching small freshwater fish and prawns in streams, and for trapping small and large fish in the sea. There were three different general types: low, circular traps with the entrance

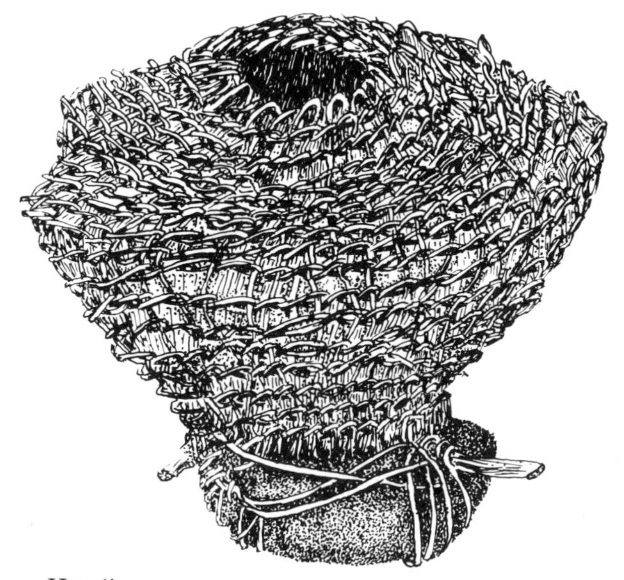

Hīnaʻi

above; long, cylindrical traps with openings at the ends; and funnel-shaped traps with small ends closed. There were permanent traps made of the aerial roots of *ʻieʻie* and temporary, roughly made ones from *ʻāwikiwiki* vines. For the *ʻieʻie* traps the bark was removed from the aerial roots and the core used, either whole or split, as warps and wefts in twining the permanent type. Leaves were stripped from vines, or not, before twining the temporary traps. Twining was facilitated by first soaking the roots for the permanent traps, and by using fresh vines for the temporary traps.

Low circular traps. Low circular traps were usually small; they were used to trap *kala* and *palani,* two surgeonfishes. These traps had a funnel or open cylinder projecting from the upper rim, with which it was continuous in structure, downward inside the trap body proper. An opening was also made in the bottom of the trap to remove the catch. The nature of the upper opening permitted fish to enter but not to leave. The twining of these traps started at the inner, lower end of the funnel, reduced gradually to the rim to make the opening, and then expanded into the body of the trap.

When these traps were in use, the bottom opening was closed by pushing two sticks through the interspace under its rim, across the opening, at right angles to each other, with the four ends projecting beyond the rim. A stone was lashed to the projecting portions of the sticks to further close the opening and to hold down the trap. Small traps were baited with crushed prawns or crabs and used by women to catch *ʻoʻopu* in streams. Women used the same kind of trap, with the same bait, to catch small *hīnālea* fish in shallow sea waters. The women walked out on calm sunny days, at low tide, and set the traps in small sandy openings in stands of coral.

Fishermen used larger traps of this type, with sweet potatoes as bait, to catch a triggerfish, *ʻuīʻuī* or *ʻuwīʻuwī;* a surgeonfish, *palani;* parrotfish, *uhu;* goatfish, *kūmū;* amberjack, *kāhala;* and squid; the fish were all deep-water inhabitants. The trap was made of upright sticks *(ʻaukā)* made of *lama* wood, which served as warps and to strengthen the trap, and *ʻieʻie* aerial roots, used for wefts. Such traps were further fitted for use in deep water by attaching two crossed sticks that arched over the opening and were fastened to the sides of the trap body. These sticks also provided attachment for a line. Sometimes a piece of the beach morning glory *(pōhuhue),* with its leaves, was twisted around part of the arch for shade. Bait for *kala* fish included the seaweed with the same name, ripe breadfruit, and half-roasted sweet potatoes.

Bait, tied on a cord, was placed in the trap, which was then lowered on a line long enough to reach the sea bottom, with a float tied to its upper end. Bait was renewed each day for four or five days. At the end of that period, the trap was drawn up and the bait examined. From the teeth marks on the bait, fishermen knew what fish were in the area. A decoy trap *(ʻapi)* was then lowered; this was also made of sticks and *ʻieʻie* aerial roots and had a wide mouth. The decoy trap was baited, and the bait was renewed each day for several days, which caused the fish to stay in the *'api*. After one day of nonbaiting the actual trap was lowered with fresh bait. After waiting for part of the day the fisherman pulled up the trap with the captured fish.

LONG CYLINDRICAL TRAPS. The low circular traps just described were used to catch a great variety of fish, but the long cylindrical traps *(hīna'i 'o'opu)* were made to catch only *'o'opu* in freshwater streams. Although not exactly cylindrical, this descriptive term was used to distinguish the trap from the low circular traps. It had a funnel opening at one end and a simple opening at the other that was not closed during use. The technique of making this trap was similar to that used in making the circular trap in that the twining began at the inner opening of the funnel. Split, peeled *'ie'ie* aerial roots were used to form fairly wide warps but the weft plies consisted of single split lengths of *'ie'ie.*

Hīna'i 'o'opu

This trap was used in this manner: the trap was set lengthwise in a main freshwater stream, without bait or cover, and the trap left in place. Because it is the nature of *'o'opu* to enter any kind of vessel to rest, they entered the trap. The fisherwoman, since it was women who used this trap, would return after a time, place her hand over the opening opposite the funnel, lift the trap, and empty the contents.

FUNNEL-SHAPED TRAPS. Funnel-shaped traps *(hīna'i 'ōpae)* were used to catch shrimps *('ōpae)* in freshwater streams; they were made from both split and unsplit *'ie'ie* aerial roots to form a large outer funnel with a tubular portion with a closed end. These traps were attended by women who used them in mountain streams. A woman, advancing in a crouching position, moved along in the stream, pulling aside small stones and thrusting sticks under larger stones to drive shrimps from their hiding places, where grass or branches of trees drooped over onto the water; the woman then scooped them up in the trap. Then she removed the shrimps from the trap and kept them in a small-mouthed gourd attached to a cord that she kept floating after her. Ferns placed just inside the mouth of the container kept lively shrimps from escaping.

Fishhooks

Fishhooks *(makau)* showed a wide range of sizes and shapes because of the great variety of fish caught with them. Fishhooks in general were divided into simple hooks, made of one piece of material, and composite hooks made of two pieces of the same or different materials, joined by lashing. Materials used were shell, human and dog bones, turtle shell, and wood. Only those made wholly or partly of wood will be described. Wood, because of its very nature, could not compete with the other

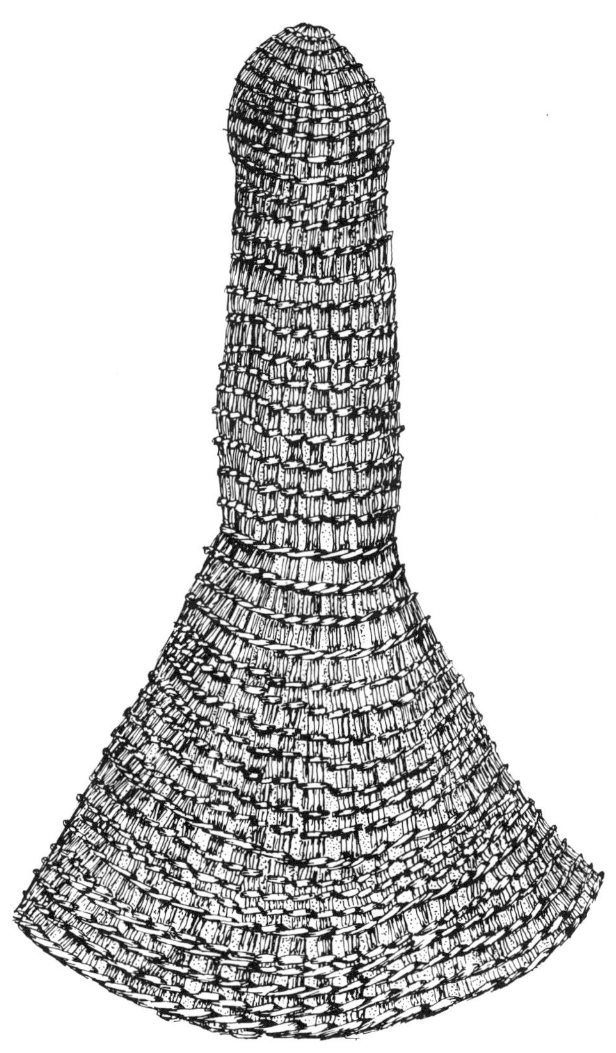

Hīnaʻi ʻōpae

materials, which were much harder, but it served its purpose in the composite hooks.

Composite hooks. *Makau manō* (shark hooks) are examples of the prime use of wood in composite fishhooks. These were made from such woods as *uhiuhi, alaheʻe, koaiʻa,* and *ʻāheahea (ʻāweoweo)* and were fitted with bone points. The hooks have two limbs connected by a U-shaped bend; they are very beautiful.

The bone points have a true upper point; this was

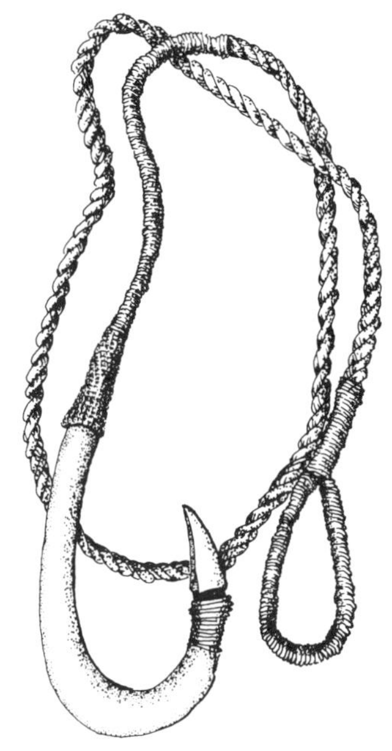

Makau manō

the part that "hooked" the shark. They are triangular in cross section and have a lower tang, a projecting part, that fits into a groove cut into the outer side of the upper end of the shank, the wooden portion of the hook. A shoulder was formed at the place where the wooden shank joins the bone tang. The wooden shank was prepared for lashing as follows: as already noted, a concave cut was made on one side of its upper end to receive the tang, and its upper rim was trimmed down to form a kind of "bed" for the point lashing. Below this bed was a raised transverse flange on the outer side to prevent the tang from slipping up under the lashing. The convex side of the tang was then fitted into the groove of the shank and lashed with *olonā* cordage.

Snood lashing. A more complicated type of point-to-shank attachment was accomplished as follows: bare shank ends had rather wide and deep grooves cut transversally across them. Long strips of *olonā* inner bark were laid longitudinally over the grooves, with the middle of the strip passing over the lowest groove. As each of these strips was attached, *olonā* cordage was used as binding thread to tightly tie down the wide strips into the grooves beneath. Then another wide strip was laid down beside the first and the tying down repeated. This process was continued until the entire end of the shank was covered. The process was completed by using binding threads over the wide strips on top of the other two grooves, and further binding the whole with threads tied at the top and bottom of the band portion of the shank.

This process was followed by another in which the lower halves of the wide *olonā* strips were doubled back over the first binding. This second layer of strips was then bound tightly tranversely over the grooves, resulting in a two-layer covering. The free ends of both layers of wide strips furnished fiber for the plies of the snood line. This process would have sufficed for a strong attachment, but the aesthetic fishhook craftsman, wishing to hide what he considered the "crude" appearance of this lashing, covered it with a snood, a textile cover, which is truly beautiful. The weft and warp were both twisted *olonā* cordage. The grooved-shank type of attachment and the textile covering, the snood, are peculiar to Hawai'i.

Fish Lines

Fish lines were made from the fibers of the inner bark (bast) of *olonā;* this cordage was strong, long-lasting, and it did not kink. The first two properties were enhanced by soaking the fibers in an infusion of the inner bark from the *kukui* tree trunk, which is rich in tannin. Futhermore, the *kukui* dye, a reddish brown, made the cordage less visible to fish.

Fishing Accessories

Fishing accessories, other than the sinkers already described, were hook and line containers and mortars and pestles for preparing bait.

Both hook and line containers were generally made from gourds and occasionally from wood.

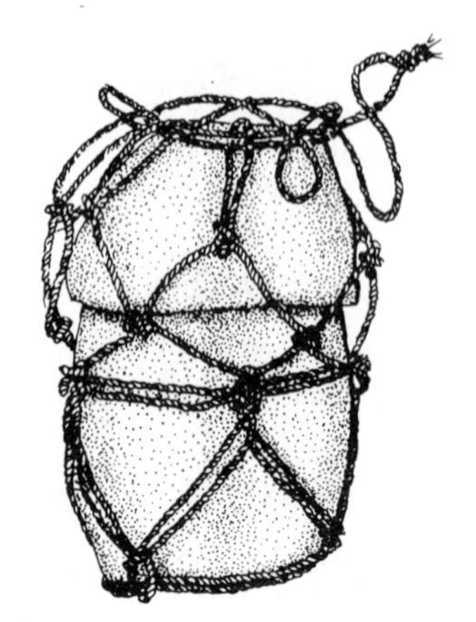

Waihona makau

Small gourds were usually used for hooks and large ones for lines; intermediate-sized gourds were used to store both lines and hooks. All these containers were fitted with covers, attached with different methods.

HOOK CONTAINERS *(waihona makau).* Hook containers were of two types: one with both the container itself and the cover made of gourds, and the other bowl of wood, *kou* for example, with a cover made from a gourd. The covers were fastened with a short net surrounding and fixed to holes in the gourd or to the upper band of a support, *'aha hāwele,* a simple net made of sennit.

In the second type of container the covers were either low or high. There were five ways of fixing covers in this second type. In one, the cover was fastened with a single piece of *olonā* twine; in another, the covered container was carried in a twine bag; in still another, *kōkō* were used; and in the others short nets were attached in one of two ways. All the fixings were made with sennit except for the last two kinds, which were made from *olonā* cordage. The technique of fixation of net fasteners for covers of hook containers was peculiar to Hawai'i.

LINE CONTAINERS. Line containers were of three shapes: globular with neck, water-bottle form, and elongated. The covers for all these were usually made of gourds, but might also be made from a coconut shell, and were attached with a single loop or cord and occasionally with a double loop.

BAIT

Four types of bait were used: hook, live, ground, and squid. A bait for hook-and-line fishing was usually small fish found in the shallow water between shore and reef, usually caught beforehand with nets.

Live bait was usually used in connection with bonito fishing.

Ground bait (chum) usually consisted of specific fish cut into pieces or pounded until soft. This bait was wrapped in the fabriclike sheath at the base of coconut fronds. A quantity of this bait was placed on a piece of this material; the latter was folded over the bait along with a hook at the end of a line. A stone was placed on the outside of the "package," and several turns of the line were made around the package and stone. A slip knot was made in the last turn of the line. This weighted bundle was carefully lowered over the side of the canoe; it sank to the bottom of the sea in the selected fishing ground. Then it was lifted a short distance and the line jerked free. The chum diffused through the water with the baited hook in the middle. Fish, attracted by the chum, eventually swallowed the hook.

The ink sac was removed from a squid, wrapped in ti leaves, and roasted on coals. A few drops were added of sap or juice from leaves, roots, flowers, or fruit from one or more of fifteen different plants; each fisherman had his favorite—and secret—selection. This liquid was mixed with the cooked ink sac and a little salt and the whole worked into a paste in a stone mortar with a stone pestle. This bait was applied to the tip of a hook; as many as sixteen different kinds of fish were caught with this type of bait.

Special mortars were used to prepare fish bait; basalt rock, either the vesicular or close-grained, was used. They were usually circular, small, and shallow. Bait mortars were also made from small coconut shells, with the distal or pointed end cut off to form an evenly rounded bottom. Simple wooden pestles were used with both the stone and coconut-shell mortars.

Bait Sticks

Bait sticks were made from such hard heavy woods as *kauila, koai'e, 'a'ali'i,* or *pua (olopua).* They were usually cut from branches from which the bark had been removed. One end was rounded, the other trimmed to a point with a very small terminal rounded knob. In shape they resembled the plummet type of stone sinker already mentioned. After completion the bait stick was charred over fire, and each time before using it was rubbed with *kukui*-nut or coconut oil. A piece of the fibrous sheath at the base of the coconut frond was used to wrap pounded *kukui*-nut kernels and coconut "meat"; this bundle was tied to the stick, which was let down into the sea by a lowering line attached at the constriction forming the neck between the stick proper and the small terminal knob. As fish were attracted to the stick they were scooped up in a net.

CATCHING OCTOPUS

Octopus *(he'e),* commonly known also as squid, were caught by one of three methods: spearing, the *kilo* method, and the use of a cowrie-shell lure.

Light spears were made of such hard woods as *alahe'e, 'ūlei,* and *uhiuhi.* Octupuses are found in the shallows and in and about rocky ledges; both men and women went to these habitats by canoes. Men speared octopuses as they lay with tentacles spread out; when speared, an octopus twines its tentacles around a spear and comes out of the hole in which it has been hiding. The fisherman killed it by biting it between the eyes or stabbing it. He strung his catch on a cord and laid the full cord on the sea bottom as he gathered more. As the tide came in, fishermen collected their catches and assembled on shore, where the whole catch was divided, sometimes as many as fifty or a hundred going to each man.

The *kilo* method was so called because octopus "fishing" was carried out where fishermen could *kilo* (see bottom). Their lure consisted of a wooden stem or spindle with a hook, a stone, and a tuft of ti leaves. The stone was circular, flat, and waterworn. One flat surface of the stone was laid against the wooden spindle, toward the front end, and tied on with *olonā* cordage. From the front end the cord was passed back over the stone to a middle line and passed around the spindle stem at the back of the stone. Longitudinal turns were made backward and forward, diagonally, to the opposite sides of the spindle. After several turns, the lashing was tightened by two longitudinal turns made on the underside of the stone, passing around the front and back ends of the lashing as they passed between stone and spindle.

The hook was a bone point without holes for attachment but with a backward extension of its base; the hook was attached to the stem with a lashing thread made of *olonā* passed around the hook extension and the wooden spindle. A cord, also of *olonā,* was tied to the front end of the spindle; when the lure was in use this cord was tied to a line.

The fisherman paddled out to a likely spot, where he chewed some *kukui*-nut meat and spat it on the surface of the sea; the oil released smoothed the water surface and he was able to see to the bottom, *(kilo).* When an octopus burrow was found, the fisherman lowered his lure about a yard away from the burrow. When the octopus saw the strange stone, its tentacles crept toward it, its body came out of its burrow, and it approached the stone until it was directly over it. The fisherman gave a jerk to the line and the octopus was hooked.

The cowrie-shell lure *(lūhe'e)* was a refined, more elaborate version of the stone lure just described. It was composed of five parts: a wooden stem or spindle, a stone sinker, a cowrie shell, a hook, and a hackle or "tail."

The sinker was an elliptical-shaped stone carved to have an inner flat surface and an outer rounded surface with a longitudinal groove in the median line of the longest diameter. The flat surface of the sinker was laid against the wooden spindle, a prepared branch of an unspecified tree longer than the sinker, at its front end, and was bound to it with *olonā* cordage with several turns backward and forward in the sinker groove, around the spindle and behind the sinker. Then a couple of turns were made around the front and back of the lashing, between the sinker and spindle, with the last turns tightening the lashing. This was a method of lashing similar to the one used in making the other lure just described.

The craftsmen who made lures chose their cowrie shells with great care; there were many varieties to choose from, and different cowries were used for different times of the day. The complete cowrie shell was used. Single holes were drilled at the front, the long sloping end, and at the back, in the middle line a short distance above the natural indentations in the lip. In the simplest and most frequently used method, a lifting cord was attached to the shell by pushing the end of the cord through the back hole from the inside of the shell and knotting it on the outside with an overhand knot; the knot was larger than the hole, so it fixed the cord securely at the back of the shell. The cord was then passed directly forward from the back hole to the front hole, with the cord lying inside of the shell between the two holes. At the inner side of the front hole a short loop of the cord was pushed up through that hole to make

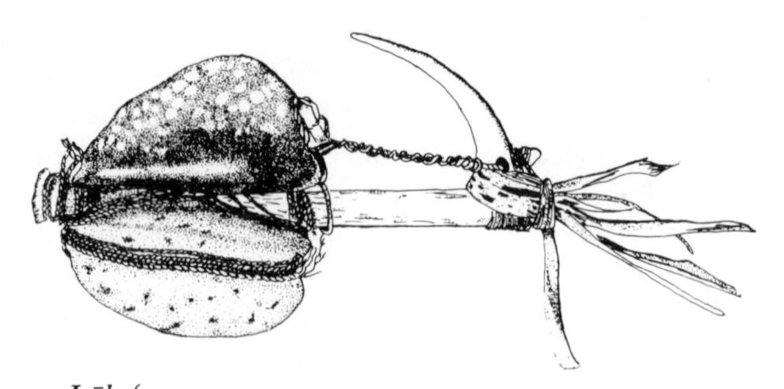

Lūhe'e

a short loop. The loose end of the loop, inside the shell, was brought outside through the front indentation of the shell, passed through the loop on the outside of the front hole, and the slack drawn taut. Thus the lifting cord was fixed to the shell, with its long end extending upward from the front hole.

The narrow opening of the shell was laid against the front of the spindle, opposite the sinker. It was tied in front with twine that made vertical turns around the spindle below the junction of the lifting cord to the hole above. At the back, it was tied in a similar manner with vertical turns around the spindle below and the fixing knot of the lifting cord. The lashings were simple and so allowed a quick change of shells.

The hook was a curved point made from dog or human bone; it was attached near the end of the spindle on the same side as the shell, with the curve pointing forward. Attachment was made by making several turns through a hole in the base of the point and around the spindle, which served as a shank for the hook. Some turns of the thread were carried forward from the point hole, around the back lashing of the shell and finally finished off with some spiral turns around the longitudinal lashing. This lashing anchored the point and prevented it from sliding back over the end of the spindle.

The hackle was a cluster of narrow strips of a ti leaf that was doubled around the front of the point, below the point hole, brought back on the short end of the spindle that extended beyond the shell, and lashed to the spindle, with several turns of thread passing around the strips above and the spindle below. The ends of the leaf strips spread out for six inches.

POND FISHING

This chapter on fishing should not be ended without writing something about the cultivation of fish in ponds *(loko)*. There were freshwater ponds *(loko wai)*, for which excavations were made specifically to be used for this purpose, or sometimes *lo'i* were used; and there were the shore ponds *(loko kuapā)*. The latter consisted of stone-wall enclosures built out from the shore both in shallow and deep water. The walls were provided with sluice gates made of mountain-apple wood. These ponds were works of art; some of them covered areas as large as sixty or seventy acres. Ponds were stocked with many kinds of fish, both by placing fry caught elsewhere in them, or by opening the sluice gates at high tide. These saltwater ponds as well as the freshwater ponds served as "fattening pens" and provided readily accessible sources of fish.

SOURCES

Beckley (1883); Bishop (1940, 41–45); Bryan (1938); Buck (1957, 285–363); Emory (1965a); Fornander (1918–1919, 118–120, 172–191); Handy and Pukui (1972, 35); Jordan and Evermann (1973); Kamakau (1976, 47–50, 59–81, 117–118); Kirch (1985, 5, 199–214); Malo (1951, 45–47, 78–79, 208–213); Maunupau (1965); Mitchell (1972, 108–115); Simmons (1963, 32–33); Stokes (1912) (1921).

4
CANOES

The Polynesians arrived in Hawai'i in canoes made from planks lashed with sennit. Once they entered the forests and saw the magnificent *koa* trees, plank canoes were a thing of the past. Although the ancient Hawaiians made large canoes from drift logs that floated down from the American Northwest, *koa* was the wood from which they preferred to make their beautiful treasured canoes.

Although *koa* was found on all the islands except Ni'ihau and Kaho'olawe, canoe logs from certain areas were considered more favorable than others, not only because of size and quality but also for the convenience of getting the logs to the shore. Such areas were the Hilo and Kona districts on the island of Hawai'i, and Hana on Maui.

The master canoe builder *(kahuna kālai wa'a)* not only knew how to select for and build a canoe, but he was also well versed in the multitudinous prayers and rituals involved at every step of canoe making. When he went into the high forests to select a tree, he might be accompanied by the person for whom he was to make the canoe, or the latter might not go to the site until the shaped log was ready to be hauled down to the shore. Other craftsmen accompanied him, carrying offerings of pig, red fish, and black fish; provisions; and adzes. After the master had made his choice, the party slept overnight near the selected tree. In the morning the food offerings were cooked in an *imu,* and more prayers said. Then it was time to cut down the tree.

The master made two parallel circumferential cuts around the base of the tree, about three feet apart, and then removed the intervening bark and some wood. This shallow cut was then deepened until, with some last strokes of the adze, the tree fell. The tree had been cut so as to fall in the direction of the main branch and clear of other trees.

Then the men waited for the appearance of Lea, the female deity of canoe builders; her visible representation was the little native bird *'elepaio.* If she walked or hopped the entire length of the felled tree without stopping, the wood was sound and could be used for a canoe, but if the bird stopped and pecked at the bark this meant that it had been invaded with insects; it was flawed and unsound. Thus the tree was condemned by the highest authority, a goddess. However, if the bird pecked only one side of the trunk it could still be used because that side could be the one hollowed. Another bird that might cause anxiety to the canoe builder was the native crow, *'alalā;* to hear its harsh cry indicated an inauspicious day for cutting a *koa* tree for a canoe.

When the tree was down and Lea had shown her approval, the master, wearing a ceremonial white tapa loincloth *(malo),* walked the length of the trunk, offering different prayers depending on the kind and future use of the canoe. When he reached the branched crown he tied a piece of *'ie'ie* vine around that part of the trunk where it was to be cut by his fellow craftsmen, and then said more prayers. The leafy crown was then cut off and the trunk debarked.

To lighten the log some preliminary trimming and

hewing was done. It was tapered at each end to form the pointed bow and broader stern, the sides were trimmed down, and the bottom was rounded. The log was then turned over and the upper side cut more or less flat. Next came a preliminary hollowing out from the top side, an operation very carefully supervised by the master; he was the one to decide where projections were to be left alongside the inner rim for seat and outrigger supports. Also, a neck *(maku'u)* was cut; this projected outward from the stern stem for attachment of the hauling line. The stern end of a canoe was made wider and deeper than the bow, so the former was cut from the base of the log.

Accompanying the men who did the trimming and hewing were two groups of men who kept the adzes sharpened: one group unloosened the lashing cords, the others did the actual sharpening. At no time, here in the forest, or later while finishing the canoe, was the inside of the canoe burned out as it was in some parts of the world.

Then it was time to haul the partially prepared log down to the sea. Although there are remnants of cleared and stone-paved runways on which to drag the logs, it is more likely, because trees were felled at fairly great distances from one another, that such logs were hauled over different routes.

Although much of the excess wood had been trimmed away the log was still a very heavy object to haul down the mountainside. Men had done the work of felling and preliminary trimming; now women and children and other men were called upon to help in the arduous task of hauling. The number of days it took to haul the semifinished hull depended on the number of people taking part. Trees and rocks that lay in the path of descent had to be cleared away. It took time and much care to lower the hull over cliffs. On the day the "dragging" began, there were great festivities among those taking part, with much food for both participants and the gods.

A great rope *(kaula ho)* made from the combined inner and outer barks of *hau,* or sennit, was used for pulling the semiworked log. At the forward end of the hauling line two more ropes of lesser diameter were attached. To each of these was lashed a stick with the ends of the lashing rope formed into a loop. Each of the two men operating these ropes thrust one shoulder through his loop. These two men kept the hauling rope straight and taut.

As the semiworked log was dragged along, the master walked about sixty feet behind it. No one could walk behind him as this space was reserved for the various canoe-building gods and goddesses.

Finally the partially worked hull was brought to the canoe shed *(hālau wa'a),* beside the sea. This usually consisted of a thatched roof on uprights, normally open on all sides. Sometimes the semiworked log was seasoned (cured) in the forest; if this had been done, finishing the canoe could now be started. If not, curing took place in the shed for months or even years. A rope of sennit was tied across the entrance of the shed to announce that it was *kapu* to all except the canoe craftsmen.

The finishing work started with the outside of the hull, with builders further shaping and smoothing the sides with adzes *(ko'i holu)* to the final thickness desired. Then the hull was turned over and the bottom trimmed, without a keel, shaping and "fairing" it *(ka'aoki)* to produce a true flowing curve. The hull was then turned over again and the hollowing of the interior finished. The narrowing of the hold—the diminishing width of the interior toward each end—made it impossible any longer to use the side-bladed hafted adze that had been used to this point, and another was substituted. This was a special socketed or swivel adze *(ko'i 'āwili,* also called *kūpā 'ai ke'e,* or simply *kūpā).* With this tool the inner sides of the hull were trimmed fairly straight from top to bottom but curved toward the bilge (the bottom of the hull).

The outer surface of the hull was finally smoothed down with a series of six different special coral and stone rubbers called *puna* and *pōhaku 'ānai,* respectively. The rubbers were more or less shaped, with the most common form a half sphere. The domed portion was for gripping; the flat side was the rubbing surface.

The hull might be painted *(pā'ele),* either before or after the accessories were added. In either case the ingredients of the paint or stain consisted of the sap of various plants: *'akoko,* the inner bark of the *kukui* root, and the flowers and buds of the banana inflorescence. The sap of the root of ti and/or that of *'uhaloa* was also sometimes added. Probably various craftsmen had their own formulas, adding other plant saps to the first-named basic three. To this concoction was added finely powdered charcoal made by burning *wiliwili* wood or *lau hala.* The paint was sometimes smeared on by hand but more often was applied with a brush made from the end of an aerial root of *hala*—one end was beaten to free the fibers from the soft interfibrous material; the other end was left intact and served as a handle.

The hull accessories consisted of a gunwale, *pale;* additional planks to build up the sides of the canoe; end pieces, *kupe;* U-shaped spreaders, *wae;* and seats, *nohona wa'a.*

Gunwales were always made of a light-colored or yellowish wood; *'ahakea* was the most favored. Others included *hōlei, 'ōhi'a lehua, 'aiea, kukui, kōlea, hō'awa,* and *'āla'a.* In more recent times, using present-day woods, whether a gunwale is present or not, the hull proper is painted black and the upper rim yellow to simulate the color combination of an ancient canoe.

The gunwale was sometimes made of one piece but more often of two or even three pieces. It was fastened to the hull proper in a peculiarly Hawaiian manner, with the lashing cord passed through unique holes, *puka 'aha,* which were remarkable longitudinal slits. A corresponding hole was made through the rim of the hull proper. The lashing cordage was flat five-ply sennit braid, pressed in three rounds through each set of holes. Wooden clamps *(pūki'i)* held down the gunwale while it was being lashed to the hull.

The end pieces, usually made from the wood of the breadfruit tree, helped prevent the entrance of seawater into the canoe; they were specifically designed planks for each side of the fore and aft sections of the hull. The end pieces for the bow and stern were alike except for two extra holes made along the upper edge of the bow for lashing to it an intermediate plank to further prevent salt water from entering the canoe. The two pieces forming an end piece were lightly lashed together with sennit. The fore and aft end pieces were then lashed to the hull in the same manner as the gunwale was lashed to the hull proper. Convex expansions of the bow point of each plank were lashed together to form an elliptical ornament *(manu),* with a similar structure at the stern end.

Thwarts or seats were made of *koa, kukui,* or breadfruit wood, shaped to snuggly fit the width of the hull; they were lashed to the cleats *(pepeiao)* on which they rested. Besides providing seats for the paddlers they served as cross braces to keep the hull from twisting and warping.

The spreaders were curved in a U shape, or at a more obtuse angle, to fit against the inner side of the hull. They were usually made from the roots of *'ōhi'a lehua* trees because of their strength and often natural curve. The roots were further shaped to fit against the inner sides of the gunwale, with the top end of each cut straight to be level with the top of the gunwale. The spreader served as a purchase point for the lashing of the booms and as an additional brace for the hull. A spreader was lashed to a pair of cleats on either side of the hull, each of which had two holes, by passing several sennit braid rounds through cleat holes, over the spreader limb, and then through the cleat hole on the other side.

The Hawaiian canoe was usually long—fifty feet and sometimes seventy or even more—in proportion to its one- or two-foot width and three-foot depth. Because of these characteristics a canoe could not have stayed upright in the water without some kind of counterbalance. In the single canoe *(wa'a kaukahi)*

Wa'a kaukahi

this was provided by the outrigger made with two booms *('iako)* and a float *(ama)*.

The booms were made of stout poles of *hau,* a light wood when dry. In most cases naturally curved branches were selected, primarily from hot arid areas where the wood was denser and therefore stronger. Sometimes branches were trained or were heated to bend them into the desired shape. The part to rest on the gunwale was cut flat on the underside to rest there solidly. Booms for the largest canoes were trimmed to make four sides.

The bark was removed from the piece of wood selected for a boom and the wood was soaked for one or two weeks in salt water, probably to prevent rot and the invasion of insects.

Booms were always attached with the outboard ends projecting on the port side, with only a short projection on the starboard side. The outer lengths were either straight or curved slightly upward before they curved downward to connect with the float at the water line. The inboard section *(kua 'iako)* was lashed to the U-shaped spreaders and the gunwale strakes on each side of the canoe. The aft boom was always closer to the stern than the fore boom was to the bow.

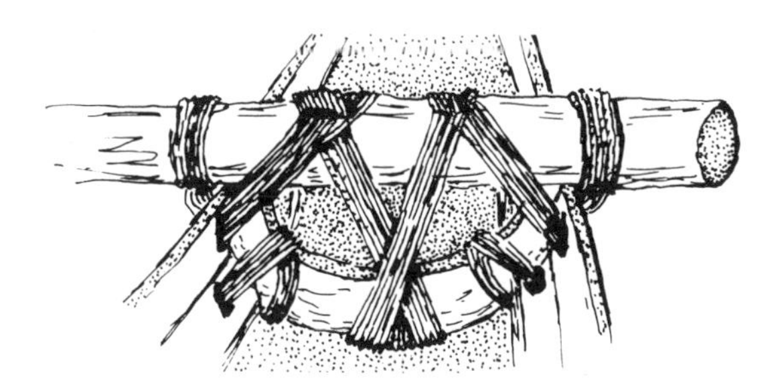

Lashing of *'iako* to *ka'ele* of single canoe

The preferred wood for the float was *wiliwili* because of its lightness; however, if that was not available *hau* was used. A well-shaped trunk or branch with a convex curve was selected. The two ends rose above the surface of the sea to minimize drag from the float. The sides were cut away from the fore end to form a vertical band. The curved form of the float as well as this specialized shape of the fore end were peculiar to Hawai'i. The aft end was usually cut off square. All units of the outrigger and its attachment to the hull were lashed with sennit with the great skill attributed to the Hawaiians.

Outrigger canoes were built for various uses, with fishing being one of the most important. Some of the fishing canoes were fitted with racks *(haka)* to hold fish spears and canoe poles; these racks were attached to the outboard part of the fore boom. The usual form was arched with a straight upper edge that was cut into three or four concave notches in which the spears and/or poles lay. The lower end was cut at a slant to lie closely, lengthwise, on the boom to which it was lashed.

A plaited mat cover *(pā'ū)* kept sea water from the hold of a canoe. The ordinary mat covered only the

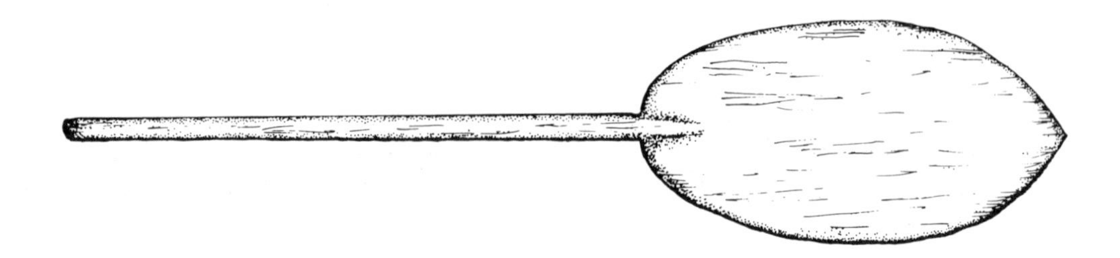

Hoe

waist (the central portion) of the canoe where materials being transported were stored. In stormy weather a cover large enough to shield the entire top of the hull was used. In this case holes were made through which paddlers fit their heads and part of their bodies. The mat was kept in place by lashing a sennit line along its sides to holes in the upper edge of the canoe, with another line crisscrossing from side to side through loops of the original line.

Hawaiian paddles *(hoe),* made of *koa,* were characterized by a straight, thick shaft *(kū'au)* and a short, wide, ovate blade *(laulau).* The blade was flat on the back surface and slightly convex, from side to side, on the front surface. The shaft continued into the blade for a distance of one or two inches to form a short midrib; this gradually decreased in thickness to its end, where it merged with the main surface of the blade. Occasionally, a short rib *(io),* extending upward from the tip of the blade, was made on the forward surface. The purpose of both ribs was perhaps to strengthen the blade or even to aid in steering.

Steering paddles were made similar to those just described but were larger, and the free end of the shaft was sometimes finished off with a straight crutch-shaped hand grip.

The Hawaiian canoe bailer *(kā wa'a)* was made from a necked gourd cut lengthwise. It is probable that the availability of gourds in Hawai'i accounts for their use instead of the usual Polynesian wooden scoop.

Anchors *(pōhaku hekau)* were made of natural unshaped stones, weighing more than ten pounds, with a drilled hole for sennit attachment. Fishermen did not often use an anchor, preferring to keep their canoes in approximately the same position by using a paddle in their left hand to steady the canoe while holding the fishing line in their right hand.

The sail *(lā)* was triangular (known as a "crab claw" sail) and made of plaited *lau hala* panels sewn together horizontally. The apex was fastened near the base of the mast. The mast *(kia),* erected perpendicularly, although positioned in various places in the canoe, was most frequently fixed beside the forward boom, either stepped on top of the boom or in front of it in the hull bottom, in a socket. One side of the sail was tied at intervals to the mast. The outer side was attached to a slender spar that functioned as a boom sprit, supporting and spreading the sail. The lower end of the spar was tied to the mast near its foot, and the upper end was drawn in toward the mast with a cord. The drawing in of this cord caused the slender boom sprit to curve inward, resulting in the free upper margin of the sail forming its unique crescentic curve. The mast was kept in position with three ropes, a back stay and a pair of shrouds, extended from the mast to each side of the canoe. Streamers of tapa were flown from the upper end of the boom sprit and sometimes from the mast head. When the sail was not in use, the mast and sail were laid flat in the canoe.

The double canoes *(wa'a kaulua)* of Hawai'i were magnificent vessels. The second canoe served as a counterbalance for the first, the role of the outrigger in the single canoe. The double canoe was normally used for long journeys and in war.

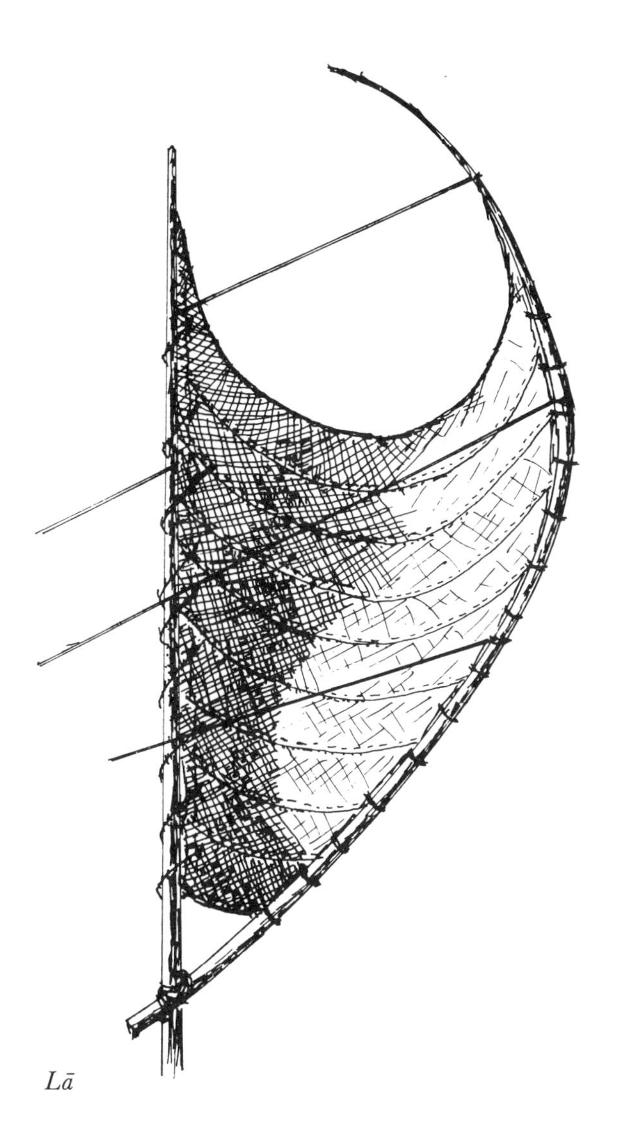

Lā

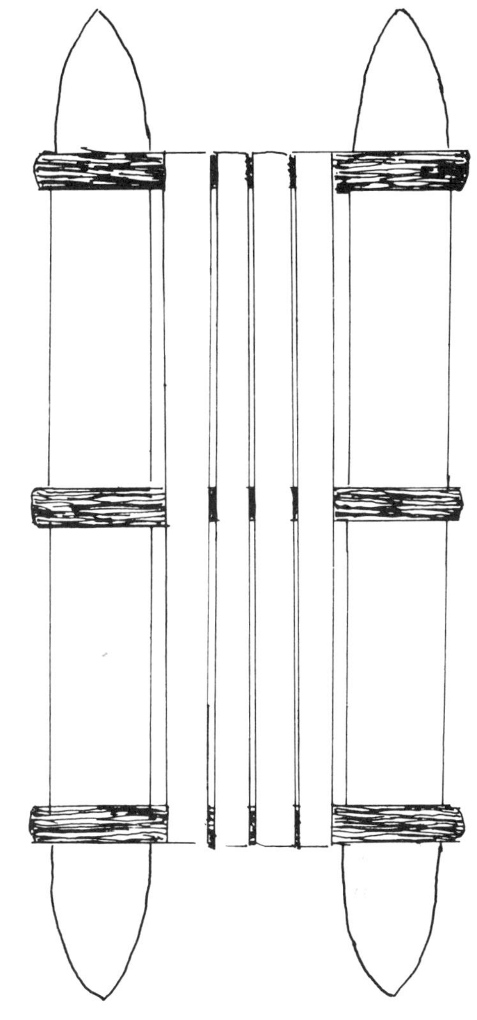

Wa'a kaulua diagram

Double canoes had two parallel hulls that were kept the required distance apart by usually more than two cross booms. The hulls were made and fitted as for a single canoe. The starboard hull was called *'ākea,* the port hull *ama,* as this was the same side as the *ama* float in the outrigger canoe. If possible, the two hulls were made of equal length; otherwise the shorter hull was placed on the port side.

The cross booms connecting the two hulls rested on the gunwale strakes, projecting a short distance on the outboard side of each hull. Although straight booms may have been used, curved booms were preferred because they raised the height of the platform *(pola)* that was built on them, thus keeping passengers and cargo dry and minimizing wave resistance. Tree branches with a curve in the center and straight ends to fit over the hulls were selected. The booms were lashed to the hulls with sennit braid.

A platform made of poles was laid lengthwise over the booms and lashed to them with sennit; over these poles, planks were laid to make a floor. For long journeys a thatched house *(hale lanalana)* was built for shelter on the platform; pens for animals taken along for food were also built on the platform. Sails of the same type as those used in the outrigger canoe were used in double canoes. When wind was insufficient, both types of canoes were paddled.

SOURCES

Bishop (1940, 45); Bryan (1938); Buck (1957, 253–284); Ellis (1969, 340–341); Finney (1977); Fornander (1918–1919, 144–147); (1919, 610–615, 618–621, 630–637); Holmes (1981); Kamakau (1976, 118–122); Kane (1971?); Kirch (1985, 6); Malo (1951, 126–135); Mitchell (1972, 116–124); Stall (1953, 29).

5
HOUSES

ALTHOUGH *kaiāulu* is the word for "village," it is perhaps more accurate to speak of a group of houses in old Hawai'i as a "dispersed community," a concept that better fits the existence in the *ahupua'a,* the land division stretching from the mountains to the sea.

The term *homestead (kauhale)* will be used to designate the group of houses belonging to a family. The location of a *kauhale* depended upon terrain as well as proximity to the source of water for the farmer, to the sea for those who fished, etc.

The main house, called *hale noa,* was the sleeping quarters for the entire family. The *(hale) mua* was the men's eating house and was *kapu* to women. There the men ate and kept and worked on their tools, and it was there the family shrine was kept. The *hale 'aina* was the eating house of women; it was *kapu* to men. Women were required to retire to a special house called *hale pe'a* during their menstrual period. If the family gods were not kept in the *(hale) mua* they were placed in the *hale heiau.* There was a food-storage shed, without sides and ends, used to store foodstuffs; this was called *hale papa'a* or *hale hoāhu.* There was also the *hale kuku* or *hale kua* where tapa was beaten. A shelter built with a roof and stone walls or plaited walls only on the windward side served as a cookhouse; it was called *hale imu* or *hale kāhumu.* It sheltered two *imu,* one to cook the men's food, the other the women's food. Firewood and cooking utensils were also stored in this shed. In those house complexes near the sea, where fisher families lived, there was added the *hālau,* the canoe shed, with its thatched roof and open sides. Canoes, fishnets, and other articles used in fishing and canoeing were kept there. The separate structures, each serving a different purpose, equate to the rooms in a single house today.

HOUSE CONSTRUCTION

Scholars have proposed that the Polynesians lived in simple lean-tos and caves when they first arrived in Hawai'i. Captain Charles Clerke (who accompanied Captain James Cook) mentions in his journal "cliff dwellings" (i.e., caves) on the steep north cliffs above Kealakekua Bay, with scaffolding and small wooden doors and ladders leading to them. These, however, may have been burial caves. When the Hawaiians were engaged in fishing expeditions or needed to gather plants in the mountains, they did occupy caves or prepare simple temporary shelters. But for permanent living there was the well-constructed thatched house; they brought the knowledge of making these from their homeland.

Houses in general were rectangular and consisted of a single room. Height and rank went hand in hand; the higher the rank of a chief the higher the ridge. But structurally, they were much the same except that material, especially the thatching material, differed somewhat with the type of house to be built and the availability of materials.

The simplest house was one without walls; the roof was built directly on or to the ground instead of being supported on vertical walls. Such houses required less material, less skill, and less labor. These houses were thatched in the same manner as the more elaborate houses, to be described later.

In areas where there was an abundance of loose stones, houses with stone walls were often built. It was a simple matter to arrange walls and then to place a thatched roof to rest on top.

Houses with gabled roofs *(hākala)* were the type in general use at the time of the arrival of the first foreigners. In this house the elevation of the roof was attained by means of a vertical framework of wood. The joining of timbers and their lashing were both peculiar to Hawaiʻi. Although the simpler houses referred to were probably built by the owner, the more elaborate ones described below were built by trained craftsmen.

Usually the floor was laid down first; it consisted of a platform of stones made without mortar. Hawaiians have been recognized as skillful "dry" masons. The next step was to cut trees to be used for posts, rafters, and purlins. These were forest trees; the ones most frequently used for the posts were such hard woods as *uhiuhi, naio, ʻaʻaliʻi, māmane,* and *olopua (pua).* Except for the more simple houses of the commoners, the bark was removed from trunks and branches. The posts were cut for lashing limbs together in a manner peculiar to Hawaiʻi. This lashing was ingenious and pleasing to the eye. Lashing cordage was usually three-ply braid made from *ʻukiʻuki,* the lily, and less often from sennit.

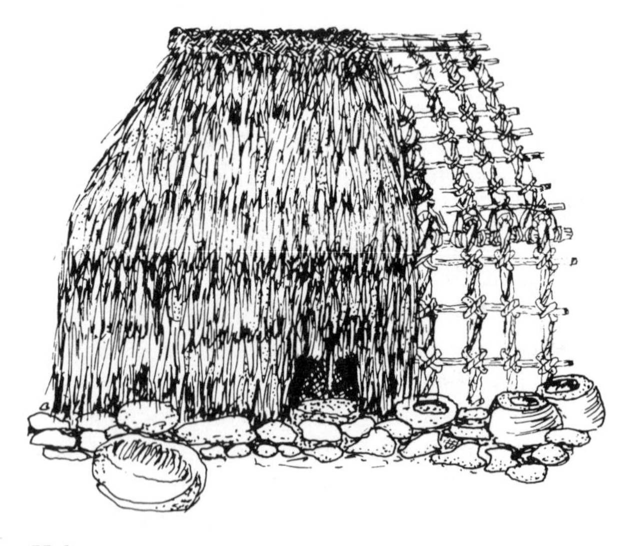

Hale

Posts

The two back corner posts *(pou kihi)* were the first posts to be erected; either these were embedded in a stone platform already laid or the platform was built surrounding the uprights after they had all been erected.

After the two back corner posts had been placed, the front ones were added. Next, the wall posts of a smaller diameter but of the same height were placed at regular intervals between the corner posts; numbers varied with the length of the house. The upper ends of corner and wall posts were notched. A square notch, open laterally, was cut; of the two flanges resulting, the inner one was short and trimmed straight at the upper edge and the outer one was prolonged upward into a pointed projection technically termed a tenon. On the outer side of a post a curved notch was cut a short distance below the base of the tenon; this was to help secure the lashing. Into these notches was fitted a horizontal pole of lesser diameter to form a wall plate *(kaola* or *lohelau).* These were lashed to corner and wall posts. The same operation was repeated for the front wall foundation.

Next, the main or median ridge posts *(pou hana),* one at each end, were put in place halfway between the corner posts; these were higher than the corner posts, determining the height of the gable. In a large house, an inner scaffold was erected to facilitate the erection of the roof assemblage. The upper ends of the median ridge posts were cut concave to receive the ends of the ridge pole, which was longer than the house length so that when lashed it had an upward bend in the middle. The lashing of the ridge pole to the ridge posts was also peculiar to Hawai'i.

Rafters

The placement of rafters *(o'a)* came next. These were usually straight poles with diameters less than those of the wall and ridge posts. The rafter length depended upon the width of the house and the slope of the roof; they were erected in pairs, one member reaching from the ridge pole to the top of the wall plate of the back wall, the other to the wall plate in front of the house. A pair of rafters ended on opposite wall posts. The ends of these rafters were cut to form a fork, with the undersides cut back as far as the beginning of the fork to form a "chin." A deep transverse groove was cut on the outer surface to keep lashing from slipping. After rafters had been laid and lashed, a second ridge pole was laid over the first, over the rafter ends, and lashed in place.

Only after the rafters were in place were the gable posts *(kukuna)* positioned, between the corner posts and median ridge post. The gable posts varied in number depending on the width of the house. They were of the same diameter as the main, first ridge post. Their height varied according to their position along the slope of the roof, being higher near the ridge post and lower near the corner posts. The upper ends of gable posts were cut obliquely to fit against the underside of rafters, which they supported. With all the upright posts installed, the wall plates in place, and placement of the rafters completed, the basic structure of the house had been completed. What remained were construction of elements onto which the thatching material would be fastened; the finishing of the interior; and the installation of a door. Ancient Hawaiian houses had no windows, thus keeping out the cold and providing greater security.

Purlins

The horizontal elements, forming a wooden framework, are technically called purlins (*'aho* in Hawaiian). There were two types: main purlins *('aho pueo)* and thatch purlins, simply called *'aho.* Main purlins were spaced horizontally on the walls and rafters and attached directly to the rafters with a simple lashing technique to form an elliptical pattern. The function of these main purlins was to brace and steady the wall and roof frameworks. Purlins needed to be the same diameter throughout their length; branches or saplings collected to be used for these items gradually decreased in diameter from base to tip, so a purlin element needed to be spliced.

The thatch or "floating" purlins consisted of horizontal elements and vertical support rods laced to each other and to the main or "fixed" purlins. Fastening was by means of running clove hitches, starting at the ridge and working down. If the elements of both kinds of purlins seemed too far apart after the task was finished, additional elements were added as needed.

Door

When erecting the basic house framework, an opening was left on one side for a doorway. In simple houses this was simply covered with a mat to close the house. In larger, more elaborate houses, such as the one just described, a doorway was constructed after thatching was completed; this consisted of two horizontal boards, one at the top and the other at the bottom, joined by two vertical boards. These were fastened together by means of grooves and drilled holes into which wooden pegs were fitted. The door frame *(kikihi)* and the door itself were preferably made from *'ahakea* wood, but wood from breadfruit and other trees was also used. The door was characterized by an arched piece set above the opening itself, suggesting the crescent

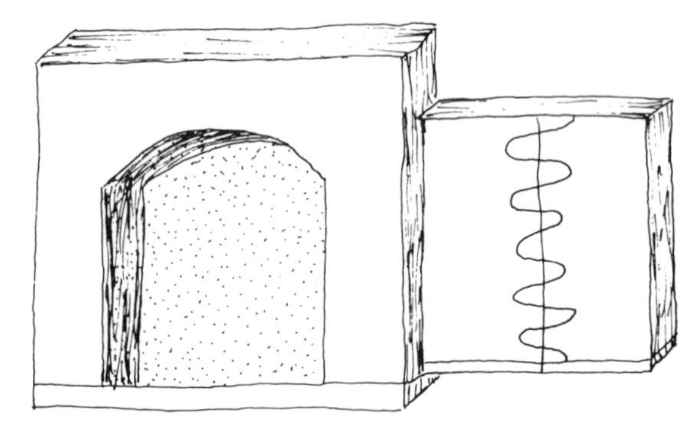

Kikihi with *puka uai*

moon. The door itself *(puka uai)* was not hinged but was a sliding one; a grooved piece of wood above the opening, on the inside, furnished the means for the door to be slid shut or open.

Thatching

After the attachment of the purlins the house was ready to be thatched. The ancient Hawaiian thatched with several different materials, but *pili* was the one preferred. The whole plant was pulled up, including roots from which soil was shaken. Thatching was started at the bottom of the front wall; the thatching "handfull," made up of several plants, with root ends up and leaf tips down, was pushed well against a purlin at the level of the stone floor platform. Usually two men worked together as a team, the one on the outside placing bundles in position and handling the tying cord, and the man on the inside cutting off any loose grass and seeing to it that the grass lay smoothly on the inside.

Braided cordage made from a lily, *'uki 'uki,* was used to tie on *pili* bundles. This was done in an interesting manner: braid was tied to the left end of the second thatch purlin from the floor, then a large bunch of *pili* was grasped by the outside thatcher, laid vertically against the purlin, and the braid passed over the outer surface of the bunch toward the right, looped over the purlin, and then hitched over itself with a couple of twists. With his left hand, the thatcher pounded the bunch to flatten it out evenly while with his right hand he pulled the braid taut. The ingenious thing about this type of thatching attachment was that making the twists in the hitch brought the braid out from the purlin, resulting in a smooth surface of the thatch rather than a "scalloped" surface as it would have been if the braid had just been looped over the purlin and pulled taut.

When the first, lowest horizontal row had been completed (the thatchers having worked completely around the house), the next row above was commenced so that there was an overlap of the first row by the second. Lashing rows were continued up the sides of the walls on to the roof, ending near the second ridge poles. The narrow space left between the two edges of the thatch here was called *ke'ehi;* the closing of this, which finished the ridge, is known as "bonneting." Bonneting was accomplished in several ways, all unique to Hawai'i.

In one method, the last rows of grass from the front and back thatching were reinforced with more *pili* held down with battens laid over it along the ridge poles and the whole finished off with a strip of braided *pili.* Another method was to lay the overlapping bases of banana leaves over the opening.

Plaited "bands" similar to those placed on the ridge were sometimes attached to the corners of the house and around the doorway as a decorative touch.

Other grasses and sedges such as *kualohia* (unidentified), the reedlike *he'upueo,* and *kāwelu,* were also occasionally used for thatching in the same manner as *pili.*

Although *lau hala* was used at times as an outside thatch, it was used primarily as a thatch to finish off houses on the inside to hide the purlin elements. Prepared *lau hala* was folded over the purlins and when two or three purlins had been covered in this manner, a cord was run along the top of the leaves to hold them flat. Sugarcane leaves were used in the same manner. Two or three thicknesses of long strips of fibers extracted from the ensheathing bases of banana were also sometimes used for this inside thatching.

After *pili,* ti leaves were probably the plant material most used for outside thatching. Only dead leaves that had fallen off plants naturally and usually dried further after falling were used. These were soaked in fresh water to make them more flexible. The leaves were then made into bundles called *pe'a.* These were tied together at their bases with another ti leaf and the blades spread out like a fan. Two leaves were bent to provide ties that were wound around the purlin in the thatching operation. These bundles were tied on with the blades hanging downward; as thatching progressed upward each *pe'a* was laid directly above the base of the *pe'a* below. With this type of thatching interior thatching was unnecessary because neither leaf bases nor purlins showed. In dry areas ti thatching is said to have lasted up to forty years; in rainy areas it did not last as long as *pili.*

The hipped-roof house differed from the gable-roof one in that the triangular portion at the ends above the wall plates sloped inward and upward instead of being vertical. In the gable roof this triangular part formed the upper part of the end wall, but in the hipped-roof house it became part of the roof. Also, in the gable-roof house the ridge pole was the same length as the walls, but in the hipped-roof house the ridge pole was shortened at each end. There were other differences but these were not great.

HOUSE FURNISHINGS

Sleeping houses were furnished very simply. Over the stone floor foundation was spread grass or old sweet-potato vines, and over this were laid plaited *lau hala* mats. Beautiful mats plaited from the sedge *makaloa* were used as sleeping mats. Both types of mats are described in the section on plaiting in chapter 2. Square-cornered pillows were plaited from *lau hala;* these were hard and were used as the Japanese use theirs, with the neck resting on one side. Bed covers were made from white tapa, several layers sewn together, with the outer, top sheet dyed and decorated. There might be a low bench to accommodate calabashes to hold wearing apparel. Kitchen utensils in the cookhouse are described in a separate chapter.

Light in the house came from two sources: stone lamps and *kukui*-nut candles. The first consisted of shaped stone vessels filled with oil from either the *kukui* or *kamani* nut, with a strip of tapa used for a wick. *Kukui*-nut candles were prepared as follows: nuts were roasted lightly, the hard shells cracked, the whole kernels removed and then threaded on lengths of dried midribs of coconut leaflets or on a splinter of bamboo. Children were usually assigned to attend to these candles. The uppermost of a string of these kernels was lighted; this would burn for two or three minutes. When it was almost burned out, the candle was inverted and the next kernel ignited. Then the burned-out nut was pushed off with a stick. The light was fairly bright but smokey. A group of candles was burned together to increase the amount of light; usually these were leaned against a stone lamp, or the latter was filled with sand and the candle ends were stuck upright in that.

SOURCES

Apple (1971); Bishop (1940, 7–9, 18); Buck (1957, 75–109); Ellis (1969, 320–322); Fornander (1918–1919, 58–65, 76–83), (1919, 58–65, 76–83); Handy and Pukui (1972, 5–14, 112–115); Kamakau (1976, 95–108); Kirch (1985, 248–257); Malo (1951, 27–30, 78, 118–126); Mitchell (1972, 157–170); Summers (1988).

6
WEARING APPAREL

THE CLIMATE of Hawai'i allowed for a minimum of clothing. The farmer wore only a so-called G-string while at work; for "dress-up" occasions this was more elaborate: wider, dyed, and decorated. The women were bare-breasted but wore a skirt similar to a sarong. Young children wore nothing. Royalty, on occasion, wore special clothing, cloaks and capes—actually regalia. Hawaiians went barefoot except as they traveled over sharp lava or coral, when they wore sandals. They wore no hats, but royalty wore helmets to complement their capes and cloaks.

Although we usually think of tapa as being the only material used for clothing by the ancient Hawaiians, early garments were also made from the fibers in banana leaf sheaths and the sedge *makaloa.* These were plaited, with the smooth plaited portion on the inside and the leaf blades or fibers hanging outside.

However, because tapa, made from bark, was the main and ultimate material used for clothing, only the making of this is described here. The Polynesians who came to Hawai'i brought the knowledge of making tapa and the primary plant from which they made it with them. Men grew, harvested, and prepared the *wauke* plant, but it was women who made and decorated tapa.

KAPA (Tapa)

Always there is the legend. The Hawaiian legend about the origin of *wauke,* the plant used most frequently for making tapa, has Maikoha, a deified hairy man, as the god of tapa makers. From his grave in Kaupō, on the island of Maui, grew the first *wauke* plant. The place of origin varies with the version of the legend; probably the origin of *wauke* in that Hawaiian locale represents the introduction of this plant to that particular area. According to this legend, Maikoha had two daughters: Lauhuki, who beat the first tapa and became the patron of tapa makers, and La'ahana, who first decorated tapa.

Bark cloth, as it is also called, is made in tropical countries around the world, but that made in Polynesia was superior to that made in any other area. And within Polynesia, Hawaiian tapa was recognized as being of the finest quality, with the greatest diversity in texture and decorative design.

Tapa was also made from the bark of the tender shoots of breadfruit trees, as in Tahiti; from *māmaki,* a close Hawaiian relative of *wauke; mā'auea* (unidentified); *ma'aloa;* and *po'o mai'a,* the upper part of banana trunks. However, *wauke* became and remained the preferred plant for tapa.

Growing *Wauke*

Wauke is a plant that requires a great deal of moisture; it was most frequently planted at the lower fringes of the wet forests. However, it was also grown, to a lesser extent, along streams and on moist bottom lands. Wherever the location, the plants were protected from winds with enclosures made of dried banana leaves until plants were hardy enough.

Land for planting was prepared by spreading plant material such as leaves over the ground and allowing this to rot; the decomposed organic matter served as a mulch, added and retained moisture to and in the soil, and served as organic fertilizer.

The preferred planting material *([huli] wauke)* was the shoots growing from the root crown left after harvesting the previous crop ("second growth"). In prying out the remainder of the shoot stem and its attached roots, great care was taken to protect the roots from injury or from breaking off. Cuttings—unrooted shoots—and pieces of roots without attached shoots were also used but were less successful as planting material.

All leaves were pinched off the planting piece, leaving only the terminal buds. Bundles of these shoots were wrapped in ti or banana leaves and tied. If water was available, the bundles were placed in it; otherwise, they were kept moist in a shady place. In both cases, the shoots were planted the next day. Before planting—February and March were the preferred months to plant—a hole for each shoot was dug in the prepared field and the soil in the hole made fine and soft with the hand.

After leaves began to sprout and additional roots formed, the field was weeded and the plants mulched. Close planting or plucking off buds of lateral branches ensured desirable straight and unbranched stems. Once the plants were growing well, the area was declared *kapu* and was not entered again until harvest time.

It usually took eighteen months or even two years to obtain the desired length and diameter. Then the stalks were carefully cut off above the ground so that the root crown was not injured. This was important because, as already stated, new shoots arising from the crown were used for planting material. Usually the leafed top was removed in the field and left there to furnish mulch and organic fertilizer.

Preparing Bark

The outer/inner bark complex can be separated from the central core rather easily. This was accomplished by making an incision in the bark along the entire length of the stem with a sharp shell, a stone knife, or split bamboo. The bark was then carefully peeled off in a process known as *pehē*. This was usually done in the following manner: the bark was loosened for a short distance on either side of the incision; the bark was rolled back from around the top of the stalk; and then, with a series of jerks, the entire bark was stripped down the entire length, all in one piece. This stripping or peeling of the bark was done either at the place of cultivation or the cut stalks were brought nearer home before stripping. The strips were then rolled into small coils, with the inner surface out, to flatten the bark.

The next step was to separate the inner bark *('i'o o loko)*—the fibrous layer or bast—from the *'ili lepo o waho,* the outer, nonfibrous layer, which was discarded. This process ("paring") was usually accomplished by scraping with a shell from *'ōlepe* (a bivalve) or from *'opihi* (a limpet), one edge of which was ground flat and sharpened, or from *una,* the shell plates of a sea turtle. Scraping was done on a board, *lā'au kahi wauke.* Woods used for this board were *'ōhi'a, kauila,* and *uhiuhi.* In an alternate method the strips were folded lengthwise, with the inner bark on the inside; the outer bark was then loosened along the fold with a bamboo knife, far enough to grab and pull off the outer bark in one piece.

Although there were several methods for preparing the bast strips to make tapa, essentially the process consisted of soaking the strips in either freshwater or seawater; giving them a preliminary beating *(hohoa)* on a flat stone *(kua pōhaku)* or wooden anvil—also used in the second stage—to form felted strips, a process known as *ho'omo'omo'o;* soaking them a second time; and then beating the mass of fibers a second time *(kuku)* on a wooden anvil *(kua kapa* or *kua kuku)* with a carved, four-sided beater.

Before going on to details about the above-named processes, the tools will be described.

Tools

The "flat stone" mentioned as having been sometimes used as an anvil in the first stage of beating is probably exactly what was used, at least in the early days; a stone was selected for shape, size, flatness, and smoothness of its upper surface or face. It is also likely that, with time, stones were specifically "dressed" for a more finished product.

The first beater *(hohoa* or *hoahoa)* was somewhat club-shaped, with a handle and a rounded striking part. The beating area sloped gradually into the handle, which decreased in diameter to a blunt point. In trimming down the handle, the edges formed were not smoothed down. The round beater separated and softened the bast fibers. There were variations of the above-described beater; one of these had a panel with longitudinal grooves forming narrow, sharp ridges *(nao),* with the remainder of the circular surface smooth. Another had panels of ridges around its entire circumference.

The second-stage beating was done on a wooden anvil called *kua kuku.* This was made from one of the following trees: *kāwaʻu* or *heaʻe,* names given to several species of an endemic rue; a native holly; a form of the native *pūkiawe;* native gardenias, known as *nānū* or *nāʻū;* and *hua lewa,* unidentified. Different woods produced different sounds; each woman had her preference.

Logs were cut to the required length and trimmed to a quadrangular section with the upper surface narrower than the lower. The ends were cut in a downward and inward slant, with a slight curve to the bottom so that the lower surface was less than the upper. A V-shaped groove, apex upward, was cut along the entire length of the undersurface and out at the ends. This wooden anvil, peculiar to Hawaiʻi, evolved from the simple trimmed-log anvil used by other Pacific people.

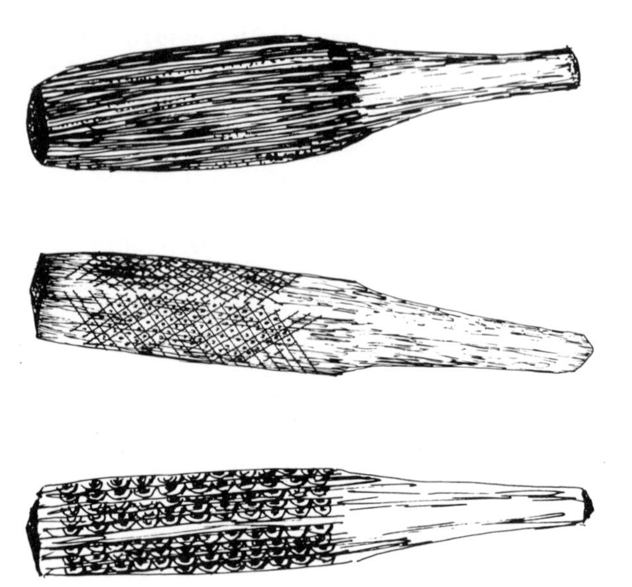

Hohoa (top) and *iʻe kuku hoʻōki (middle and bottom)*

When beaten upon, the *kua* was raised above the ground by means of *lau hala* pillows, wooden blocks, or stones placed under the curved ends. The anvil wood, together with the groove, produced an agreeable sound. It is not difficult to imagine that the ancient Hawaiian women chanted an accompani-

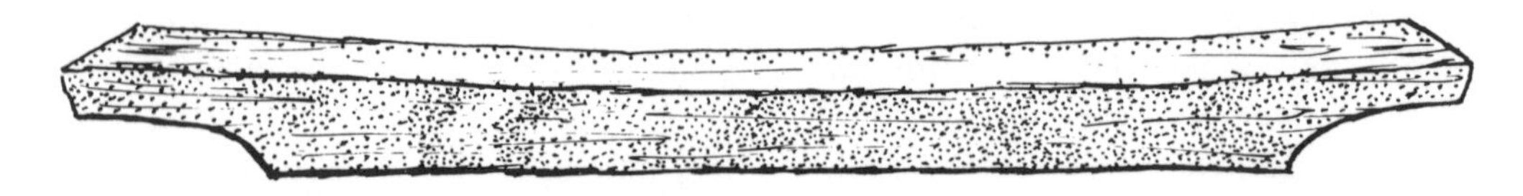

Kua kuku

ment to the sound made as they beat tapa on these anvils.

Beaters for the second-stage beating *(i'e kuku)* had a quadrangular blade with four surfaces of equal width. The handle *(kū'au)* was trimmed on the round and narrowed gradually from the junction with the blade to end in a blunt point. After the beater had been shaped, the surfaces of the blade were rubbed smooth and then long parallel grooves were carved with a shark tooth set in a handle of wood or bone, or with a splinter of sharp stone; male experts did this carving with the aid of a bamboo ruler. For the early stages in this second beating, grooves were deepened to form distinct ridges *(nao);* the furrows between these ridges were called *'au-waha.* A single beater might have each of the four surfaces with ridges of different widths—seen in sequence, one side would have ridges wide and rounded off to form inverted Us; next, sides with ridges of decreasing widths; to the fourth side with very narrow ridges, with edges like inverted Vs.

In many beaters the four surfaces were carved differently so that a single beater could be used for different purposes. One side might be left smooth *(mole),* without carving; this was used at the end of beating to produce a smooth surface on very thin tapa, or one to be printed with some design. Or widely spaced shallow grooves were sometimes cut in this *mole* surface in designs to create a watermark, unique to Hawai'i.

Patterns to create the watermark design on pre-contact beaters were simple but still attractive. These beaters were used in the finishing stage of beating and were called *i'e kuku ho'ōki.* Transverse lines were carved to cross longitudinal patterns to form a series of squares or rectangles (*mole hālu'a* pattern); or oblique lines were crossed to form lozenges. Sometimes only pits were made (*mole pūpū* pattern), or longitudinal rows of pits alternated with longitudinal lines (grooves), or pits were made in the squares formed in the *mole hālu'a* pattern; this last pattern was popular and was called *mole hālu'a-pūpū* pattern. The pits *(pūpū)* were, when shallow, dug with a mounted shark tooth; the deeper ones, with a drill pump. Postcontact watermark beaters had much more elaborate designs after the availability of metal tools.

Beaters for both stages were made from hard woods, the primary one being *koai'a.* Other woods used were *uhiuhi, kauila, lehua, o'a* (a form of *kauila* on Maui), and *nīoi.*

Processing Bark

There was a preliminary soaking of the bast strips in either salt water or fresh water; there seems not to have been a preference of the one over the other. The duration of soaking varied from a few days to one or two weeks. The purpose of the first soaking was to remove an objectionable slimy sap and to soften the fibers.

Beating the Bark

In the first beating *(ho'omo'omo'o* or *hohoa),* several soaked strips, varying from two to ten, were laid one upon the other on a stone or wooden anvil and beaten with the *hohoa,* the first-stage beater. This is a felting process; the product was called *mo'o-mo'o.* This first beating was probably done out of doors.

Mo'omo'o were set out in the sun to dry and bleach. When thoroughly dry, these strips were bundled and wrapped in a piece of tapa so that they could be stored until time for the second soaking and beating.

Before undergoing the second beating, the *mo'o-mo'o* were subjected to a second soaking, during which they underwent a fermentation process. The strips were wrapped in ti leaves and soaked in fresh running water or placed in water in a closed vessel for ten days to two weeks; or they were sprinkled with water and covered with banana leaves. The resulting product was a very soft, slimy strip that barely held its shape. This was torn, rolled, and pressed into a mass to make a kind of "cake." This resembled paper pulp but was made tough by kneading.

Although direct quotations from references have been avoided in this book, one from Makekau (1899) is presented here because that author so beautifully—and adequately—described the second beating:

> Now the wahine kuku sat down with a [calabash] of water by her side, picked up [her beater] and with two hands began to beat. She beat from left to right and back. The stuff spread out and became tougher, and to beat it evenly, she changed it from right to left, etc.—
>
> From morn to eve, the low sound of her beat could be heard. The wahine kuku beat merrily on, sprinkling water as she beat and also changing her sitting position and the side of her ie [beater]. The finished part was coming toward her and once in a while she rolled it up.
>
> When finished, she folded it in two. This was called wili-oki. When done, she beat it into one with the ie wawahi and then folded and beat the sides or edges to make them straight.
>
> When there was nothing more to do, no mistakes to be corrected, no thick places to be beaten again, no thin places to be filled and nothing done over again, her day's work was finished and a cloth, called kapa, which resembled flexible paper had been made.

As noted, the *mo'omo'o* was kept damp by sprinkling with water during this beating; sometimes *palaholo,* the sticky sap of furled *'ama'u* fern fronds, or the sap of *māmaki* was mixed with the sprinkling water.

It was because of the fermentation process and the second beating that Hawaiian tapa was of such high quality, superior to any other made in the Pacific. Whereas in tapa made in other island groups in the Pacific distinct fibers were still visible, and holes were left in the finished product, Hawaiian tapa was homogeneous and without defects of any kind.

Large pieces of tapa were obtained by overlapping the edges of smaller pieces and beating the joined portions together. The sticky sap from furled *'ama'u* fern fronds was also used to join two pieces. Breadfruit-sap "gum" *(kēpau)* or thinned *poi (kakale)* may also have been used as an adhesive.

The second-stage beating was done in a thatched shed, *hale kuku.* For a single worker this would have been small; for a greater number of workers it would have been large. Sometimes the same shed was used for dyeing and decoration. The *hale kuku* was *kapu* to men.

The first clothing made by a native people would naturally be utilitarian in character. It was not important whether the material from which it was made was left its natural color or bleached or whether or not it had adornment—all that was needed was something to cover the body. So let us assume that the first tapa made had the natural color of the plant bast used—unbleached, undyed, and undecorated. The first step away from this original material was bleaching.

Bleaching

After the second beating, tapa was laid flat upon stones or on level land, usually in specially prepared drying yards *(pā kaula'i)* enclosed with stone walls or stick fences to prevent damage to the tapa by animals. Natural stones of suitable shape, size, and weight were used to hold down tapa sheets while they were drying. These holding stones were shifted from time to time to keep the tapa flat and prevent its tearing; tapa was often turned during the drying process. Tapas were brought in when it rained, and always at night. Another method of bleaching was to spread the tapa over the moss growing on rocks in a stream.

Dyeing

Eventually natives would want something more colorful and decorative. Leaving native people in general, let us concentrate on the Hawaiians. As Polynesians, they came to Hawai'i already skilled in

DYE PLANTS AND DYES

Hawaiian name	Common name	Scientific name	Plant part used	Color obtained	
				English	**Hawaiian**
ʻaʻaliʻi	Hawaiian hopseed bush	*Dodonaea viscosa*	Capsules	Red	
ʻākala	Hawaiian raspberries	*Rubus hawaiensis* and *R. macraei*	Fruit	Pink	*ʻĀkala*
ʻalaʻala wai nui	Peperomia	*Peperomia* spp.	Leaves and stems	Grayish green	*ʻĀhiahia* or *puahia*
alaheʻe	None	*Psydrax odorata*	Leaves	Black	
ʻamaʻu	None	*Sadleria cyatheoides*	Young fronds	Red	
hame, haʻā, or *mehame*	None	*Antidesma* spp.	Fruit	Red to dark purple	
hōlei	None	*Ochrosia* spp.	Bark of stem and roots	Yellow	
kamani	Alexandrian laurel	*Calophyllum inophyllum*	Fruit husk	Brownish mauve	
kauila	None	*Alphitonia ponderosa*	Leaves and bark	Bluish	
kō	Sugarcane	*Saccaharum officinarum*	Leaves and stem	Black (from charcoal)	
koa	Koa	*Acacia koa*	Bark	Red	
kokiʻo or *hau hele ʻula*	None	*Hibiscus* spp. *Kokia* spp.	Flower petals	Pink and lavender	
kōlea	None	*Myrsine* spp.	Bark	Red	
			Stem wood	Black (from charcoal)	
kou	None	*Cordia subcordata*	Leaves (senescent)	Warm brown	
kūkaenēnē or *ʻaiakanēnē*	None	*Coprosma ernodeoides*	Stem inner bark	Yellow	
			Fruit	Purple to black	

Continued

DYE PLANTS AND DYES (continued)

Hawaiian name	**Common name**	**Scientific name**	**Plant part used**	**Color obtained**	
				English	**Hawaiian**
kukui	Candlenut tree	*Aleurites moluccana*	Fruit husk	Grayish/beige	
			Inner bark trunk	Brownish red	
			Inner bark root	Reddish brown	
ma'o	Hawaiian cotton	*Gossypium tomentosum*	Leaves	Greenish	*'Ōma'o*
			Flower petals	Yellow	
milo	Portia tree	*Thespesia populnea*	Fruit wall	Yellowish green	
nānū	Native gardenias	*Gardenia* spp.	Fruit	Yellow	*Nā'ū*
naupaka kuahiwi	Mountain *naupaka*	*Scaevola* spp.	Fruit	Purplish black	
noni	Indian mulberry	*Morinda citrifolia*	Inner bark of trunk	Yellow; red**	*Māhuna*
			Inner bark of root	Yellow; red**	
'ōhi'a 'ai	Mountain apple or Malay apple	*Syzygium malaccense*	Inner bark of trunk	Brown	
			Inner bark of root	Brown	
			Fruit skin	Red	
'ōlapa	None	*Cheirodendron* spp.	Fruit, leaves, bark	Bluish-black	
'ōlena	Turmeric	*Curcuma longa*	Rhizome (underground stem)	Yellow to gold to mustard (depending on maturity)	*Lena* or *'ōlena* and *'ōlenalena*
pā'ihi	None	*Nasturtium sarmentosum*	Bark of stem	Black	

Continued

Hawaiian name	Common name	Scientific name	Plant part used	Color obtained	
				English	Hawaiian
pala'ā	Lace fern	*Sphenomeris chinensis*	Old leaves	Brownish red	
pili	Pili grass	*Heteropogon contortus*	Blades	Black (from charcoal)	
pōpolo	Glossy nightshade	*Solanum americanum*	Fruit	Blackish purple	
			Leaves	Green	
pōpolo kū mai	Native pokeberry	*Phytolacca sandwicensis*	Fruit	Dark purple	
'uki'uki	None	*Dianella sandwicensis*	Fruit	Purple-blue, true blue**	
'ūlei	Hawaiian hawthorn	*Osteomeles anthyllidifolia*	Fruit	Lavender	
'ulu	Breadfruit	*Artocarpus altilis*	Male inflorescence	Yellow to tan to brown (depending on maturity)	

**When lime was added.

Sources: The data contained in this table come primarily from research conducted by Yamasaki (1982), from the references listed at the end of this chapter, and from research by the author.

the art of tapa dyeing and such decorative means as stenciling and painting (brush work). As Hawaiians they continued to dye their tapa and decorate as their forebears did but then abandoned these for methods they developed in their new home.

The literature on dyes and dyeing is very limited; much of the present-day knowledge on these subjects has been acquired through the work, in recent years, of persons interested in experimenting with plants named as dye plants in the old literature. This information is incorporated in this book, including the accompanying table. Nonplant dyes, which are not covered, were a clay, *'alaea,* with colors ranging through the oranges and reds; and a sea urchin, *wana (Diadema paucispinum),* which produced a blue dye.

The most common method of dyeing is by immersion; the Hawaiians called this method *ho'olu'u.* The dye itself was called *waiho'olu'u;* the experts who prepared the dyes, *po'e ho'olu'u;* and the house in which dyeing was carried on, *hale ho'olu'u,* if this operation was conducted in a separate house from the tapa-beating house.

The fresh dye-plant material was pounded or crushed in a stone mortar with a stone pestle, both made and used only for the preparation of dye materials. An infusion was made by adding water; this was strained through a piece of the fabriclike sheath at the base of a young coconut frond. This had been prepared by cleaning off the surface debris, after which it very much resembled our modern cheesecloth.

Penetration of the dye was enhanced by heating the dye; since the Hawaiians had only wooden and gourd containers, water was heated by placing hot stones in the dye in these vessels. However, the dye was heated only if it was obtained from barks, roots, or rhizomes; the proteins in leaves, flowers, and fleshy fruits, from which some dyes were obtained, are precipitated with heat, which hinders release of the dye material.

Hawaiians were also aware of the desirability of adding a mordant, called *ke kāliki (liki),* to a dye; this is a substance that, with the dye material, forms an insoluble compound in a fabric. For mordants Hawaiians used salt and urine; these are also used by native people around the world. In the literature lime (burnt limestone: calcium oxide) is listed as a mordant; this is not correct. Lime is used to change the acidity of a dye to an alkaline state and thus change the color (see table of plants used for dyeing).

The total immersion of large pieces of tapa would have required large vessels to serve as vats and great quantities of dye. However, Hawaiians did use this method to a limited extent, as well as pre-dyeing the bast before beating. Other methods of coloring entire sheets of tapa are described under painting.

Painting (Brush Work)

There is little information in the literature about Hawaiian painting; mention is only made of stenciling. Since this is a method for decoration in other Pacific islands, such as Fiji, it can be assumed that the methods used there were probably used by the Hawaiians. In Fiji, patterns were cut in hala leaves as well as in banana-leaf sheaths. Dyes and brushes for use with stencils were prepared as for lining, to be described below. The use of stencils was eventually abandoned.

In one type of painting the surface of white tapa was very lightly brushed with dye, so lightly that though the color showed up clearly on that surface it did not soak through to the other side. However, if this type of brushing was done on the surface of very thin tapa, the dye soaked through and the tapa had the appearance of being immersion-dyed. The brushing was done with a folded piece of tapa dipped in the dye.

Another method was to brush the surface of tapa with a "charcoal bag," a piece of tapa filled with powdered charcoal *(lānahu* or *nānahu)* obtained by burning sugarcane or *kukui* nuts. This produced gray tapa.

Overlaying or Bonding

Two names were used for tapa with this type of decoration: *kapa pa'ūpa'ū* and *kapa pa'i'ula.* Although these two names are sometimes used interchangeably in the literature, and the end product is somewhat the same, the methods of making them are different. For *kapa pa'ūpa'ū,* two strips of the prepared bast *(mo'omo'o),* one dyed and the other not, were beaten together so that each side and therefore each "color" remained discrete. For *kapa pa'i'ula,* pieces of dyed tapa were beaten into a white piece of tapa; sometimes as many as three different colors were "inlayed." The use of this method resulted in the color being blotchy. These tapas were used primarily if not wholly for *kilohana,* the outer sheet or layer in a set of bed coverings.

Lining

Although this is a form of painting, the subject is treated separately. The design was created by drawing series of straight or diagonal lines or stripes parallel to or at angles to one another. To add to the attractiveness of the design, the lines were painted not only in black but in colors and/or the spaces filled with color. Sometimes the lines were zigzags or a row of dots.

The lines were drawn freehand; with the aid of bamboo "rulers"; or with one of two types of tools. One type, made of *kauila* wood, had the appearance of a knife, with a round handle and a thin, sharp-pointed blade about the thickness of the handle. The upper edge of the blade was blunt; the lower was trimmed to a sharp edge, which was dipped in the dye to print a single straight line. A multiple-line liner was also made of *kauila* wood. This was made like the single-line liner except that it had a wider blade, which was divided into five evenly spaced prongs, resembling the teeth of a comb.

Besides being made out of *kauila,* the single- and multiple-line liners were also made from bamboo. Prongs in these varied in number from one to four, and in width; sometimes wide and narrow prongs alternated, or the outer prongs were wider than those between. With these different arrangements, a great variety of decorative designs was obtained.

Although one might think that such liners were used as pens, by dipping them in the dye and then drawing them along to form lines, they were actually used as printers or stamps.

Brushes for stenciling and free-hand lining were made from dry *hala* keys (drupes) from which the epidermis had been removed from the inner end to expose a tuft of fibers; these were cleaned by teasing them with a bamboo splinter. The other hard outer end served as a convenient handle. For broad lines and for stenciling, these brushes were used as described; for fine lines the fibers were trimmed.

To extract dye material from plant tissues, a considerable amount of water must be used. Although the resulting product is ideal for an immersion dye, it is too "watery" for stenciling and lining; there is "bleeding," with "blurry" designs resulting. There is no mention in the literature of how the desired thicker dye was obtained. Experimentation has shown that an acceptable viscous "paint" can be obtained by evaporating the thinner dye through exposure in a shallow vessel or by heating. The Hawaiians probably used one or the other, or perhaps both, methods; in the second method they would have placed hot stones in a calabash filled with the dye.

Four tapa designs

Cord Rubbing

Another type of design, *kaula kākau* (from *kaula,* cord, and *kākau,* to print), was originally assumed to be produced by a process known as "cord snapping," where a length of cord is dipped in dye, the two ends held taut, and the middle lifted and then allowed to snap back on the fabric. However, this was not the method used by the Hawaiians; the tapa was laid over the cord (two-ply *hau* cordage) and then the dye was rubbed on top, along the length of the cord. This produced a design showing the oblique twists of the cord.

Block Printing

Precontact printing was done with designs carved on blocks of wood; these were simple but very attractive. Later, after the introduction of metal cutting tools intricate designs were carved on bamboo and used as printers. A discussion of these is outside the scope of this book; however, it should be pointed out that although the development and execution of these latter designs were influenced by the early foreigners, they should be considered unmistakably Hawaiian.

Grooving

Although grooving or ribbing is not a form of design, it was a process in which the character of "plain" tapa was altered and another dimension added, as was the addition of the designs described above. Ribbed or grooved tapa, called *kuaʻula,* was made with a grooved board *(papa hole)* and a grooving tool *(koʻi hole).* The *papa hole* was made of *kauila* or some other hard wood. It was grooved on both sides with longitudinal parallel lines; one side had finer lines (ridges) than the other.

Although the Hawaiian word for the grooving tool refers to a pig's jawbone and there are references to this use, the more usual grooving tool was made of wood. It had a long, narrow edge, curving up at the ends. The upper border was shorter and thicker, to allow a firm handgrip.

The technique of grooving or ribbing consisted of dampening the tapa and then laying it over the grooving board in the position desired. A bamboo ruler was placed over the tapa in the long direction of the grooved board; the ruler was held down with the left hand and served as a guide for the grooving tool, held in the right hand against the ruler. The grooving tool was then rubbed back and forth to press the tapa down into a corresponding groove in the board beneath. The ruler was moved as each groove was made so that the completed ridges between grooves stood out. Grooving in tapa was evidently a means of making thick tapa more pliable, an asset with the *pāʻū* and *malo* garments, for which tight wrapping was required.

Perfuming

Hawaiians perfumed their tapa in one of two ways: either the scented material was added to the dye, before the latter's application, or it was laid between the folds of stored tapa.

The perfume plant most frequently added to a dye was the native fern *lauaʻe.* The fern fronds were first heated with hot stones and the fragrant sap pressed out; this was mixed with the oil from roasted coconut meat and added to the dye. The dye to which this perfume was added was one made from *ʻōlena,* and was reserved for *malo* worn by chiefs. Another perfume sometimes added to a dye was the sap of the *kamani* tree.

For scenting finished tapa, a number of plants were used: the powder derived from the heartwood of the native sandalwood, *ʻiliahi;* the aromatic rhizomes (underground stems) of *ʻawapuhi kuahiwi,* shampoo ginger; leis made of the anise-scented capsules and leaved twigs of *mokihana,* and of the sweet-scented branches of *maile;* the orange-blossom-scented flowers of *kamani;* and the fragrant male inflorescences of the *hala* tree.

CLOTHING

The clothing of the ancient Hawaiians included the *malo,* a loincloth, for men; the *pāʻū,* a skirt, for women; and the *kīhei,* a cape, for both men and women.

Malo

A *malo* was a long, narrow strip of tapa made from a single *mo'omo'o,* lengthened by adding other pieces at the end during the process of beating or pieces joined together by stitching. Colors for dyeing and patterns for decoration used for *malo* were different from those used for *pā'ū.* A *malo* was put on by the wearer by holding one end temporarily under the chin or between the teeth, passing the piece of tapa back between the legs, and then winding it around the waist after hooking it around the first turn at the back. Here it was tied in a knot or dropped, as was the end held in front. Thus, with the latter method, there was a flap (a little "apron") in front and another in back.

Pā'ū

A *pā'ū,* rectangular in shape, reached from the waist to the knees or lower, leaving the breasts bare. A *pā'ū* was wound around the waist and the end tucked in at the waist. Sometimes a smooth pebble was rolled up in the upper edge and tucked in to prevent the end from slipping loose. Perhaps a cord at the waist made the *pā'ū* more secure.

There was another type of *pā'ū,* which consisted of several, often five and even more, up to ten, layers of tapa. Such multilayered *pā'ū* were made by fastening the sheets together on one of the long sides with short pieces of cord made of twisted white tapa. These ties were looped through the sheets at intervals and then tied, with an extra twist, with an overhand knot. The innermost and outermost sheets were often decorated with liner-printed designs.

Kīhei

The Polynesians who came to settle in Hawai'i brought with them the wearing apparel described in the preceding paragraphs. Such clothing sufficed in the year-round mild climate of the central Pacific islands from which they came. However, the climate in Hawai'i was different—there was a rainy, chilly season *(ho'oilo)* from October to April that was uncomfortable with clothing that left the shoulders bare. So the Hawaiians added another garment to their wearing apparel, the *kīhei,* a cape. Not much is known about the *kīhei.* Although it has been generally assumed that this garment was prepared from tapa specifically for this use, any length of tapa may have been used.

SANDALS *(Kāma'a)*

The ancient Hawaiians normally went barefoot; however, when walking over rough coral or on the rough surface of *'a'ā* (lava) flows, even their tough soles needed protection. Faced with this predicament, the traveler quickly prepared adequate protection by fashioning simple sandals by folding or plaiting any tough fibrous plant material at hand: *lau hala,* ti leaves, or banana leaf bases, or the bark of *hau* or *wauke.*

The simplest sandal made from the bark of *hau* or *wauke* was made by forming a loop from a cord twisted or braided from stout strips long enough to fit the foot. One end of the loop was used to shape the sole. The sole was made by twining a second cord, the weft, over and between the two parallel sides of the loop, the warp. The loose ends of the loop cord, beyond the sole portion, were used to tie the sandal to the ankle.

More elaborate, and more comfortable, sandals were commonly made from ti leaves. There were two types, one with and one without a median strand. For the sandal without a median strand, two pairs of two leaves each were twisted and knotted to form a framework, the length of the sole of the sandal. The sole was formed by inserting leaves from opposite sides, crossing under and over the opposite marginal strands and each other, alternately. This weaving was started at a knotted end and worked toward the middle. This process was then repeated from the other end until the overlapping turns met in the middle. The distal ends of the leaves, left untwisted when making the sole framework, were

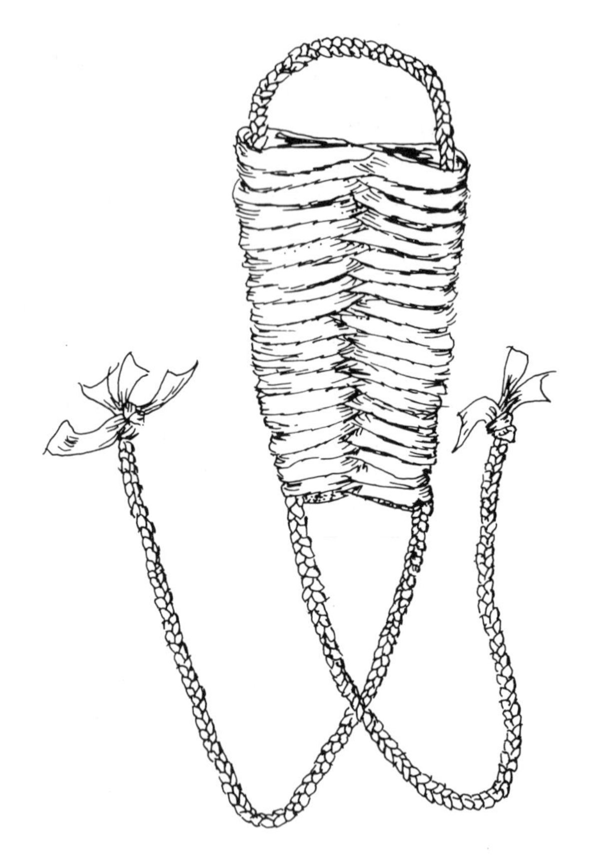

Kāma'a

twisted to make tying elements. These were pulled to tighten the woven strands of the sole and tied with an overhand knot to form two loops, one beyond each end of the sole. One of these was for slipping over the toes; the other was brought up to the first loop and tied in a knot.

In the sandal with the median strand, the frame for the sole was made the same as for the other sandal, with a middle strand added. In weaving the sole of this sandal, the weaving strands were crossed alternately around the central strand and either side. Half turns were sometimes made as the weaving proceeded so that the lower border became the upper border on the next turn; this made for a thick-soled and therefore more sturdy sandal.

RAINCOATS *(Kui lā'ī)*

Raincoats were worn primarily by the fowlers *(kia manu)* for protection from the rain and cold of the upland forests, where they went to capture the birds from which were plucked feathers to decorate the capes, cloaks, and helmets described in the next section. A raincoat was made from a rectangular piece of ordinary one-inch-mesh fishnet made of *olonā* or hibiscus cordage on which ti leaves were tied. Ti leaves, with stems attached, were split along the midrib. Then, beginning at the lower border of the netting, the half leaves were attached on one surface of the netting by passing a half stem of a half leaf around two arms of a mesh below the mesh knot and then tying the free end around the standing part of the stem with an overhand knot. A slipknot was thus formed that could be tightened by pulling the blade portion. The half leaves were tied singly along the lower border, below every second mesh knot in the same row. The leaves of this first row hung below

Kui lā'ī

the lower edge of the netting. Successive rows were attached until the entire surface was covered up to the upper, neck border. A cord was passed through the neck marginal meshes, knotted to the meshes at each end, and the cord ends left free for tying.

REGALIA

Feather Capes and Cloaks *(ʻAhu ʻula)*

The feather capes, cloaks, and helmets are included here with clothing although actually these were very special—regalia, the symbolic clothing of royalty. They are of particular beauty and magnificence and are without equal anywhere else in the world.

The general name for feather capes and cloaks *(ʻahu ʻula)* comes from *ʻahu,* a cape, and *ʻula,* red; in the beginning all or at least most capes and cloaks were made with red feathers because red was considered a "chiefly color."

Although feather capes were and are made in other parts of Polynesia (in New Zealand by the Maoris, and in Tahiti), each people developed its own technique. The main feature of the Hawaiian technique, essentially different from the others, was the preparation of a netting base and the attachment of feathers to it. The netting foundation was made of cordage prepared from *olonā.* The netting ranged from a very fine mesh for the small feathers from forest birds to a coarse mesh, like a fishnet, for large feathers from various seabirds (tropicbirds, the man-of-war hawk) and domestic fowls. The feather garments were made in various shapes—trapezoidal and circular capes, and circular cloaks.

The trapezoidal capes were of two types: one was made with coarse netting on which large feathers were tied; the second type was made from fine netting to which smaller feathers were tied. Trapezoidal capes were finished on the upper and side borders with braid made from *olonā* cordage, in some cases, up to eight ply, or with a band made from tapa or plaited *makaloa* sewn to the *olonā* binding at the upper border. In the latter case the band was plaited to the correct size with the edges turned in to pre-

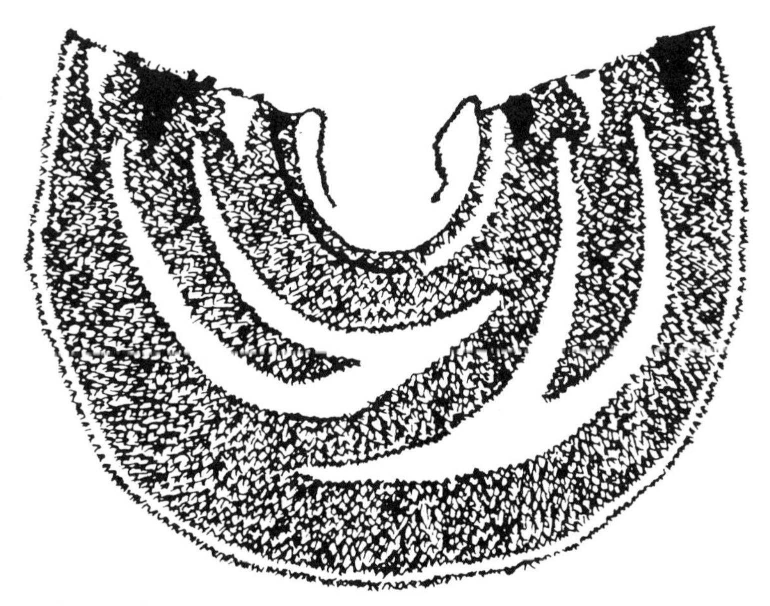

ʻAhu ʻula (cape)

vent unraveling. The braid finishing the upper border was extended beyond the sides of the net, and the free ends were then tied around the neck. Or, the side borders were sometimes extended beyond the upper border to form loops through which the arms were thrust.

The following technique of tying feathers on the netting was probably common to all the coarser type of trapezoidal capes: a single feather was laid with its quill over a mesh knot, pointing upward, and tied with fine *olonā* cordage. The feathers were arranged to form simple color designs of black horizontal bands along the lower border, or vertical bands or rectangles of black, brown, and sometimes white along the side edges, with the remaining area filled in with white. To relieve the drabness of the body of the cape, an upper border with smaller red and/or yellow feathers was added. These were too small to be tied to the large meshes of these capes, so they were first fixed on a separate band of tapa or *makaloa* matting. Sometimes pieces of bird skins, with feathers attached, were gummed to precut tapa or preplaited *makaloa* bands. These were attached in turn to the remaining base netting at the upper border, above the uppermost row of large feathers.

In the fine trapezoidal capes, with convex lower borders, bunches of smaller feathers were tied on a fine netting; this technique is described later in the discussion of the circular capes and cloaks.

Trapezoidal capes were worn by lesser chiefs, probably because chiefs of lower rank were not able to obtain enough small colored feathers to make large circular capes, and also because with time the more spectacular capes came to be the regalia reserved for superior chiefs and royalty.

The circular capes had a straight neck border, a much longer convex lower border, and two straight side edges that met in front. They were deeper in the middle line at the back than in the front. There was a great deal of variation in the length of the neck border and the depth of the curve from cape to cape.

The netting for these capes was made from a two-ply *olonā* cord with the same netting knot as used to make a fishnet. The mesh was so small that the ordinary netting shuttle could not be used; instead, a fine form of shuttle was used, consisting of a round stick of wood cut down to form a shoulder at the end of the handle part, then trimmed to a long slender prong with a blunt point. The cord was attached near the point by two half hitches. When even finer meshes were made, a piece of the midrib of a coconut leaflet was used instead for a shuttle.

Rather than making a net to the size and shape of a cape, prepared net was cut to the required size and shape. However, a cape could seldom be cut from one piece of netting. This problem was solved by making a median gore with two side pieces, or fitting together pieces of different shapes and sizes. The edges of adjoining pieces were fastened together by running a cord of *olonā* alternately through marginal meshes on each side. The upper and side edges of the cape were bound with an eight-ply braid of *olonā* cordage. The ragged cut edge of the lower border was not finished off because it would be covered by the lowermost row(s) of feathers.

The smaller feathers used in making the circular capes were fastened to the fine netting as follows. First, instead of tying on a single feather at a time, bundles or bunches of feathers, sometimes as many as eighteen, were attached. A simple knot was used, with the number of knots tied around the feather bunch varying with the direction in which the thread was moved from one bunch of feathers to the next, and with the length of the quills.

The feather craftsman tied the first row of feathers along the lower border, working from left to right, and from bottom to top, as in thatching. Because the bottom had a pronounced convex curve, all the rows were curved and parallel to the first row; this necessitated the occasional filling in with extra rows toward the center as the rows progressed from bottom to top.

Although capes were sometimes made with feathers of one color, red or yellow, more often they were

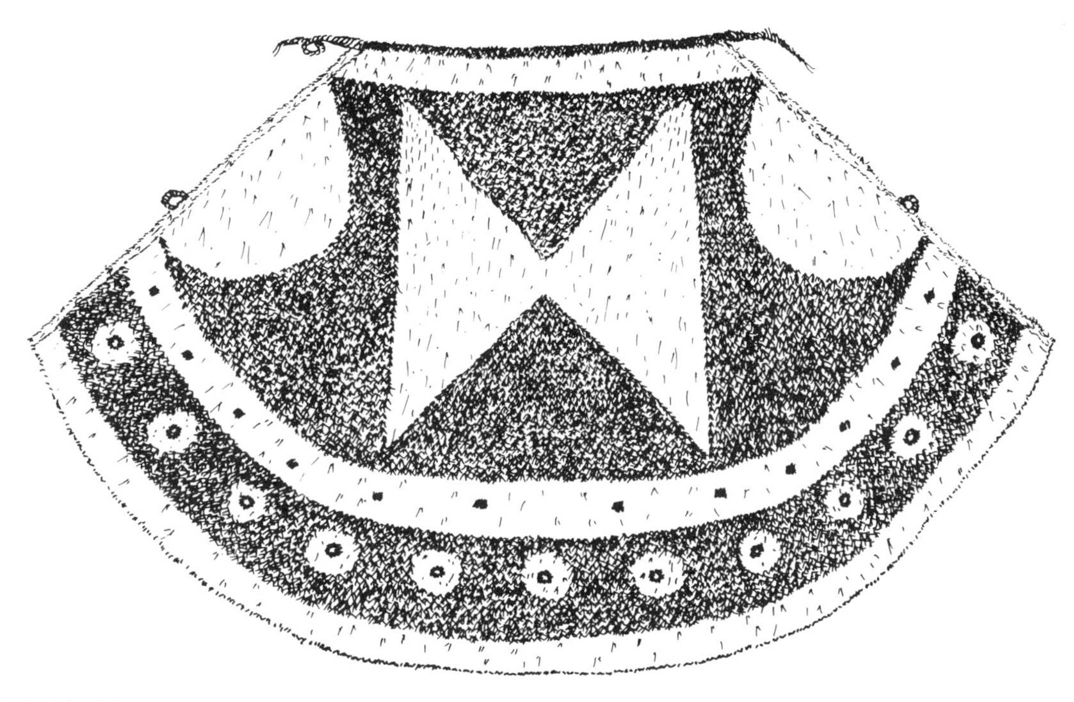

'Ahu 'ula (cloak)

decorated with feathers of a contrasting color. Black, and rarely green, feathers were used and then only in creating the motif. The most common motif was the crescent; it seems natural that the curved shape of the cape itself would affect the motif design. This crescent might be one large, centrally placed one, or a row of smaller ones. Sometimes, there would be half crescents on each of the sides of the cape. Their bases faced outward so that full crescents were formed when the two sides were drawn together. Besides crescents, lozenges and triangles were used for decorative motifs. Besides the designs noted above, there was often a border along the bottom and sometimes along the neckline, of a contrasting color, different from that in the body of the cape but usually the same color as the decorative motifs.

The cloaks were longer and wider than capes but were of the same general style and color; they usually had a red background with a yellow border and yellow motifs, with the latter commonly one or more crescents. Additional motifs introduced in the decorating of cloaks were bars, disks, and triangles.

It is outside the scope of this book to describe the birds whose feathers were used to make capes and cloaks and the helmets that are discussed next, and how the feathers were gathered and prepared.

Helmets *(Mahiole)*

The magnificent crested helmets were unique to Hawai'i and were a natural and worthy complement to the feathered cloaks. Their similarity, in shape, to the Greek helmet is often remarked upon, but there is no known connection between the two. Their shape has also been attributed by some to the influence of members of the crew of a Spanish ship that was supposedly shipwrecked on the island of Hawai'i early in the sixteenth century; however, such helmets existed in Hawai'i earlier than that. More probably the crest was made to simulate the comb of a cock, a highly prized fowl of the ancient Hawaiians.

The cap, the basis of all types of helmets and appendages, was made from the aerial roots of *'ie'ie.* Long roots were prepared for weaving or twining by peeling (removing their bark), splitting them, and then soaking them to make them pliable. These prepared roots were bound together in warps and wefts in a type of finger weaving. Feathers were attached to a fine-mesh netting, as with capes and cloaks, and the net was then fitted over the twined base.

Helmets with low median crests, covered with feathers, were the most commonly made. Crests began a short distance back from the front edge and then gradually diminished in height to end flush with the back edge of the cap. The color varied considerably; however, those used most frequently were the red of *'i'iwi* feathers on the cap and sides of the crest and the yellow feathes of *'ō'ō* on the upper surface of the crest. In some helmets, small-spaced patches of black feathers were placed along the borders of the cap, or longitudinal bands of black feathers were alternated with the red.

The wide-crested helmets had crests that were wider than but not as high as the low-crested ones. The front end of the crest of a wide-crested helmet was more often vertical in contrast to the forward-pointing projection of the low-crested helmet. Also, the crest of this helmet was probably twined separately and then attached in contrast to the twined-in-place crest of the low-crested helmet. The feathers used in the wide-crested helmets were applied by the braid method: feathers were tied to one side of a three-ply braid of coconut-husk fibers, then these braids were arranged and tied on to the cap with *olonā* thread, following the longitudinal contour. Some braid was cut short to fit the narrower back. Feathered braid differed in color; there might be alternating bands of green and red with tufts of yellow and green feathers at long intervals. Sometimes a black band was introduced into the color scheme, for contrast.

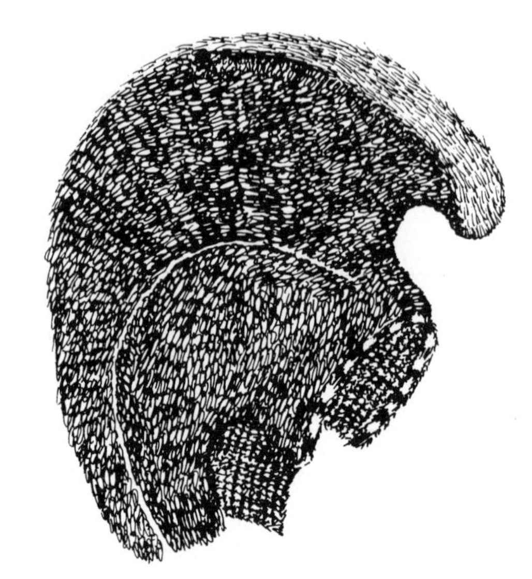

Crested *mahiole*

Crescent-crested helmets were distinguished by an elevated crescent-shaped crest connected with the cap by means of four to six props. The props were made from long hairpin warps pushed through the cap and then continued to form the crest. Props were all straight except for the anterior one, which leaned—curved forward—to accommodate the forward curve of the crescent. The height of props decreased going from front to back. The front end of the crest projected some distance beyond the anterior prop, in some helmets more than in others. A variation from the vertical props were those bent into a zigzag pattern, with the points of adjacent zigs and zags touching.

"Mushroom"-ornamented helmets had special mushroomlike ornaments in place of crests and were never feathered. The ornaments, twined separately in the form of cylindrical stems with rectangular bases and surmounted with circular disks, were then attached to the cap. The number of these ornaments varied.

Hair helmets, which might even be considered wigs, were rare; at any rate, few remain. The technique for attaching the hair to the cap was as follows: three or four tufts of bleached hair were placed

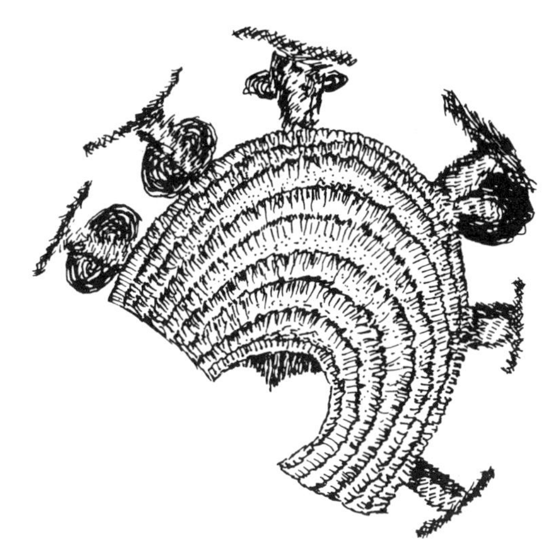

"Mushroom"-ornamented *mahiole*

together and tied at one end with *olonā* thread. The tied ends were then covered by doubling a separate tuft over them. These prepared tufts were tied with two-ply *olonā* thread to the cap in rows, starting at the back edge with the tied ends pointing forward.

ORNAMENTS *(Kāhiko)*

The Hawaiians had a greater number and richer variety of body ornaments than any other Polynesians. These were made of plant parts, feathers, shells, the teeth of sperm whales and dogs, boar tusks, turtle shells, human finger bones, and human hair. Only those made of plant parts are discussed here.

Leis

Wreaths or garlands *(lei)* for the head *(lei po'o)* and neck *(lei 'ā'ī)* either were of a temporary nature, made from fresh leaves, flowers, and fruits, or were semipermanent, made of seeds, seed capsules, and nuts.

Leis used as ornaments only are described here; those used for medicine or as offerings are treated in relevant chapters.

For temporary or perishable leis, fronds of various ferns were used: primarily the lace fern, *pala'ā; palapalai* or *palai;* and other ferns such as *laua'e* and the rare *pala.* This last fern was more often presented as an offering to the gods than worn as a garland.

Plants known as "fern allies," but more primitive than ferns, were used in leis. These were *moa* and the tropical club moss *wāwae'iole.*

Flowers used for leis included, primarily, *'ilima* and different species of *lehua,* such as *'ōhi'a lehua, lehua 'āhihi,* and *lehua papa;* candlenut, *kukui;* native gardenias, *nānū; koki'o;* the native sesbania, *'ohai; nuku 'i'iwi* or *kā 'i'iwi; hala pepe; māmane;* sugarcane, *kō;* and *'āwikiwiki* or *puakauhi.* To this list of the more commonly used flowers could probably be added many more that were less often used.

The stems of the native dodder, *kauna'oa,* a leafless parasitic vine, were twisted or twined into an orange-colored lei.

Leaves from the following plants were used either alone or with one or more of the flowers listed above: *'ōhi'a lehua,* older and young leaves *(liko);* the favorite, *maile; lā'ī,* ti leaves; *'ōlapa;* a native lily, *pa'iniu; ōhelo; pūkiawe; kukui;* and *'a'ali'i.* Leaves from these plants were also used with one or more of the following fruits: fruitlets of the screw pine, *hala;* the fruit or berries of *'ākia, 'aiakanēnē* or *kūkaenēnē, 'ōhelo,* and *'ūlei;* and seed capsules of *kāmakahala, 'a'ali'i, 'ōhi'a lehua,* and *mokihana* were also used.

One seaweed, *limu kala,* was used as a ceremonial lei.

The leis were made from the plants listed above in the following ways:

1. *Nīpu'u* or *hīpu'u:* This was a knotting method in which leaf stems or short lengths of vines were knotted together to make a lei long enough to encircle the head or to drape around the neck. An example of such a lei is one made from *maile.*
2. *Hili* or *hilo:* This was a braiding or plaiting method of a single plant material. The lei made

from the fern *pala'ā* is an example of this technique.

3. *Haku:* A lei made by this method resembled one made by the *hili* method except that flowers were added at intervals to the three-strand plait as it was being plaited.

4. *Wili:* This was a winding method; small bunches of flowers, leaves, and/or fruits were attached to a center core made from a strip of dried banana stalk "skin," *lau hala,* or a folded, "deboned" ti leaf by winding a string around the stems of the decorative plant material and the core.

5. *Kui:* In this method the plant material, flowers, and/or fruits were strung through the center or side. There were several kinds of *kui* leis. The simplest was "straight" stringing, lengthwise through the centers of flowers or fruits; a lei made in this manner was called *kui pololei.* An example of such a lei is one made of *'ilima* blossoms. An *'ilima* lei for the neck requires a thousand flowers; these are picked early in the morning when still unopened. At one time *'ilima* leis were reserved for royalty.

Another type of *kui* lei was made by round stringing; these were called *kui poepoe* and were made by stringing crosswise through the stems or bases of flowers and arranging the flowers like spokes in a wheel. In still another lei, *kui lau,* flowers were also strung crosswise but laid flat, with flowers arranged alternately, from side to side of the string. This type of garland is now called *lei maunaloa.*

6. *Humupapa* or *kui papa:* In this method the plant materials were sewn on a foundation of dried banana stalk skin, *lau hala,* or folded, deboned ti leaves.

For winding, tying, and sewing the ancient Hawaiians used the bast fibers of *hau,* although sometimes the finer fibers from aerial roots of *hala* were used, especially for the *kui ilima* leis. Needles were made from the midrib of a coconut leaflet, pointed at one end and flattened at the other end by chewing on it; an eye was made in the flattened end with a finger nail.

Perishable leis were stored and carried in one of several types of receptacles. Several types were made from ti leaves, and another type from a section of one of the overlapping leaf bases forming a banana trunk. A large lei or several leis were accommodated in a ti-leaf carrier *(pū'olo lā'ī)* made from eight to ten leaves, depending on the size of the leaves. These leaves were left on the end of their stalk and the stalk broken or cut off to a length of a foot or more below the lowest leaf. The bunch of leaves was then turned so that the stalk pointed up with the leaves spread around it on a flat surface. The lei or leis were wound around the stalk beginning at the level of the leaves. The leaves were then drawn up around the lei in rotation so that each leaf in turn overlapped the preceding leaf. When all leaves had been drawn up around the lei(s) the ti leaf tips were tied securely to the stalk, leaving several inches of stalk beyond the leaf tips for a carrying handle. A dry ti leaf was used for tying because a green leaf cracks and tears. Ti leaves were used to make several other less elaborate kinds of packets.

The banana-sheath holder for leis was made from a section cut from the trunk; each overlapping leaf base made a holder. These made a kind of tray on which a lei could be stored and/or carried. The leaf-base section has a tendency to recurl, so the container becomes a closed packet and the lei remains moist and undamaged.

Leis were made from the seed capsules of *mokihana;* these were semipermanent and could be kept for several years, during which time they retained their aniselike fragrance.

The only truly permanent leis made from plant materials were those made from seeds and nuts, *lei hua.* The seeds of several plants were used—soapberry, *a'e* or *mānele;* a member of the sapodilla family, *āulu* or *kaulu; lonomea;* and *wiliwili*—but the favorite of all this type of permanent *lei* was made from the nut of the *kukui* tree.

A *kukui* nut was prepared for stringing, for this was the manner in which it was used, by cutting off the pointed end to form one hole, and another hole was made by puncturing the other end from the inside out. The kernel, usually called "meat," was removed by breaking it into small pieces through one of the holes, or by burying the nuts in the ground. Placing them in protected areas where the kernel is consumed by ants is a postcontact method.

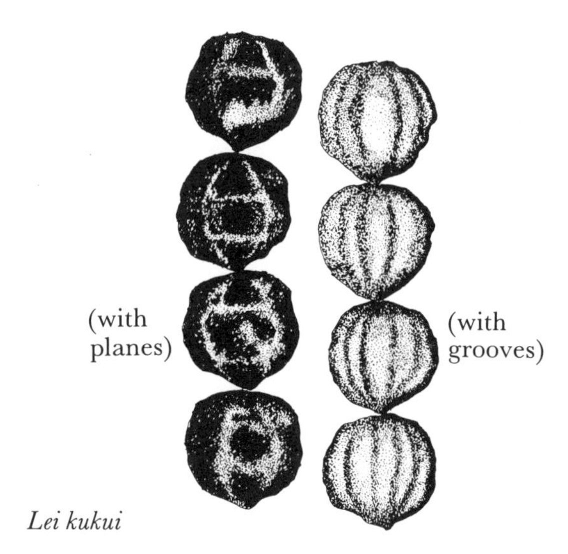

Lei kukui

The surface of the nut was finished in various ways: the original shallow furrows and shape were retained; or the surface was smoothed and artificially grooved lengthwise between the two holes, with three grooves on the naturally flat surface, three on the opposite naturally rounded side, and one at each of the two ends, the resulting seven planes separated by distinct longitudinal edges; or, rarely, elliptical planes were made on each of the narrower surfaces of the nut.

Files made from the spines of the sea urchin were used to make the grooves, and the surfaces were finished by "sanding" with the skin of a fish, *kala,* or of the shark, and finally with pumice. The nut was polished with a piece of tapa dipped in *kukui*-nut oil.

Permanent leis were wrapped in soft tapa and stored in covered calabashes (*'umeke pōhue* or *'umeke lā'au)* made of gourds and wood, respectively.

SOURCES

Bishop (1940, 23-39); Brigham (1911); Bryan (1965); Buck (1957, 165-252); Ellis (1969, 108-113); Fornander (1919, 614-619, 636-641); Handy (1940, 196-199); Handy and Pukui (1972, 12, 44, 112, 178); Ihara (1979); Kamakau (1976, 108-116); Kaeppler (1970), (1975), (1980); McDonald (1976); Malo (1951, 48-49); Makekau (1899); Mitchell (1972, 171-190); Simmons (1963, 34-35); Webb (1965); Yamasaki (1982).

7
MUSICAL INSTRUMENTS

THE ANCIENT Hawaiians were a poetic people; their poetry is among the most beautiful in the world.

Because they had no written language, their poetry was related in *oli* and *mele*. *Oli* were recited without accompaniment, predominately on a single prolonged pitch, occasionally varied by a quavering of the voice or an added pitch. *Mele* involved several pitches with the words arranged as to be intoned (chanted). Thus the *mele* somewhat resembles the song of today. Unlike the *oli*, the *mele* was accompanied by a *hula*, a musical instrument, or both. It is with the musical instruments used to accompany the *mele* that this chapter is concerned.

Did the sounds associated with plants grown by and gathered from the forest by the ancient Hawaiians influence their use for musical instruments?—the clatter of bamboo stalks in the wind, the rattling of dried seeds in an old gourd, the sharp crack of struck wood, the sound made by slapping a hollow log or gourd?

Hawaiian musical instruments, except the stone pebbles *(ʻiliʻili)* used as castanets and the conch-shell trumpet *(pū)*, were made, at least in part, from plant materials. Bamboo *(ʻohe)* was used in the largest number of instruments, with gourds *(ipu)* a close second. Materials came from the native trees *kauila*, *ʻiliahi*, *ʻūlei*, and *ʻōhiʻa lehua;* a vine, climbing pandanus, *ʻieʻie;* and a shrub, *olonā;* and the Polynesian-introduced trees coconut *(niu)* and breadfruit *(ʻulu);* and a shrub, paper mulberry *(wauke)*.

Although the rattle made from three gourds *(ʻūlili)*, the ti-leaf whistle *(pū lāʻī)*, the coconut-shell bull roarer *(oeoe)*, and the *kamani*-nut whistle are sometimes listed among and described as musical instruments, they were actually toys and are described in this book in the chapter on games and sports.

The most numerous musical instruments in the world are those in which the material they are made of produces the sound. In Hawaiʻi these included beating sticks *(kālāʻau)*, footboard treadle *(papa hehi)*, stamping tubes or bamboo pipes *(ʻohe kāʻekeʻeke)*, split-bamboo rattle *(pūʻili)*, single-gourd or single-coconut rattle *(ʻulīʻulī)*, and the double-gourd drum *(pā ipu)*.

KĀ LĀʻAU (Beating Sticks)

These so-called hula sticks were made in pairs, of unequal length, with one considerably longer than the other. *Kauila* was the wood exclusively used to make *kā lāʻau*. A dancer or musical accompanist cradled the longer stick near its center with the left hand, one section lying along the forearm, like a violin, close to the side of the body, the other section projecting in front. The shorter stick held in the right hand was tapped against the longer stick to create the rhythm for the *mele* and dance, *hula kā lāʻau*.

Footboards or treadles *(papa hehi)* were used in conjunction with these sticks. The footboards consisted of a small piece of wood with one surface convex, the other flat or slightly concave. A crosspiece under the board facilitated its movement as a treadle. The board was used to beat time with the right foot. Before the use of the footboard the performer stamped time on a flat stone lying either flat or leaning against another stone. Or the player beat time with a foot on a hollow vessel, probably a gourd, which lay inverted on the ground.

'OHE KĀ'EKE'EKE OR *PAHŪPAHŪ* (Stamping Tubes or Bamboo Pipes)

These instruments were made of lengths of bamboo that included both a node and an internode. Bamboo of different lengths and diameters produced musical notes of different pitch. The musician squatted holding one *kā'eke'eke* vertically in each hand, alternately striking the basal (closed) end of each on a firm surface such as a stone, the floor of the *hālau,* or some other firm nonresonant body or material. The thin walls and long internodes of the native *'ohe* lent themselves to a finer tone than any of the later-introduced bamboo varieties. An ensemble of musicians, each operating two pipes, furnished the accompaniment for the appropriate *mele* and *hula kā'eke'eke.*

RATTLES

Pū'ili (Split-bamboo Rattle or Fringed Bamboo Tube)

This was a musical instrument played by the performers of the *hula pū'ili* and the appropriate *mele.* It consisted of a length of bamboo including a node and an internode; usually a portion of a second internode was left that, with the node, formed the handle of the *pū'ili.*

The long section of internode was slit into narrow slats around its circumference, from the outer open edge to within a short distance of the node, the handle end. In some *pu'ili* alternate slats were removed to allow more movement to those remaining, and thus a more pleasing sound was created.

Pū'ili

The *hula pū'ili* was performed by individuals seated cross legged or kneeling in two rows, with each performer in a row facing a partner in the other row. Players simultaneously tapped their facing partners lightly on either shoulder or wrist, creating a cross-action. In some *hula pū'ili* the ground between partners was also tapped. At certain intervals the rattles were exchanged in midair.

'Ulī'ulī (Single-gourd or Single-coconut Rattle)

The prototype of the *'ulī'ulī* was a small gourd or coconut filled with small pebbles or shells and fitted with a handle. The dancer grasped the instrument by the handle and shook it or struck it against parts of the body and/or the ground to produce a rattling sound to accompany the *hula 'ulī'ulī* and the appropriate *mele.*

A gourd or coconut about the size of a large orange was prepared by drilling three or four small holes on the circumference a little below the point of stalk attachment. Water was added through the holes, the contents fermented, and the liquified contents were then emptied through the holes. The gourd or coconut shell was properly rinsed, allowed to dry, and small pebbles or shells were added through the holes.

The holes were further utilized to form the handle, which was made from either the leaves or aerial roots of *'ie'ie,* the native vine related to *hala.* When four holes were made, green leaves or the debarked softened roots were passed through two opposite holes and drawn to the same length. This process was continued until the four holes could hold no more. When three holes were made, leaves or roots were inserted into each of two holes with the ends from both drawn through the third hole. Sometimes additional leaves or roots were added to the outside of the column formed to thicken the handle. All of the drawn-through ends were bunched together and

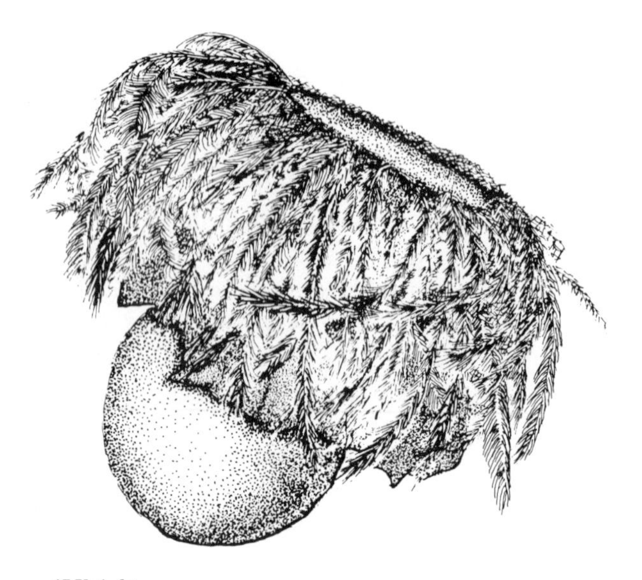

'Ulī'ulī

tied for several inches with transverse turns of sennit. The handle was thus fixed firmly to the gourd or coconut. When the leaves or roots dried, they became stiff and hard, producing a rigid handle.

This simple rattle served its purpose well. However, as with many other articles used by native peoples, more elaborate and/or decorated forms evolved from simpler ones. So it was with the *'ulī'ulī*. Instead of the fairly short ends of leaves or roots used to form the handle of the simple rattle, the ends were left long enough to spread out horizontally to create evenly spaced radials. These were fixed into position with a circular hoop made of the same plant material by looping the ends of the radials around the hoop from the inner side, and tying the doubled-back ends of the leaves or roots to their tangential radials. The resulting disk framework is best pictured by comparing it to a wheel: the bunched upper ends of the handle form the hub, the radials are similar to the spokes, and the hoop to which the latter are fastened resembles the felly or rim.

The disk framework was covered with a circular piece of tapa. Another piece of tapa covered the upper portion of the handle, extending outward to cover the under surface of the radials. For ornamentation, a fringe of cock feathers from native fowl was sewn, by their quill ends, to the top covering. The down-curving feathers hung gracefully over the edge. A second circular piece of tapa was gummed over the original top circle of tapa to hide the quills. Sometimes pieces of bird skin and feathers were gummed on the upper circle of tapa.

Pebbles or small shells were pushed through the holes before the handles were made; the shaking of this instrument produced the rattling sound for which it is named.

The modern *'ulī'ulī* is quite different from the ancient instrument just described. The body is made from the fruit of the introduced calabash tree, *la'amia;* seeds of the introduced wild, yellow-, sometimes red-flowered canna, *ali'ipoe,* are used to create the rattling sound; modern fabric is used instead of tapa for covering the disk and the underside of the disk; doweling is used for the handle; and dyed duck feathers are used to ornament the rim. Also, the dancer today holds an *'ulī'ulī* in each hand instead of a single instrument in one hand, either the right or left, as formerly.

DRUMS

Pā Ipu, Kuolo, or *Ipu Pa'i* (Double-gourd Drum)

The double-gourd drum, as the name indicates, was made of two gourds joined together one above the other. The lower gourd, the larger of the two, was usually of the long globular variety; the upper one was of the short squatty or roundish variety.

The instrument was made as follows: the stem ends of both gourds were cut off, that of the upper gourd to make an opening larger than that of the lower gourd. The two gourds were fitted together by inverting the upper gourd over the lower one. As the angle of the upper gourd was more obtuse, the rim of the lower gourd's opening projected inside the

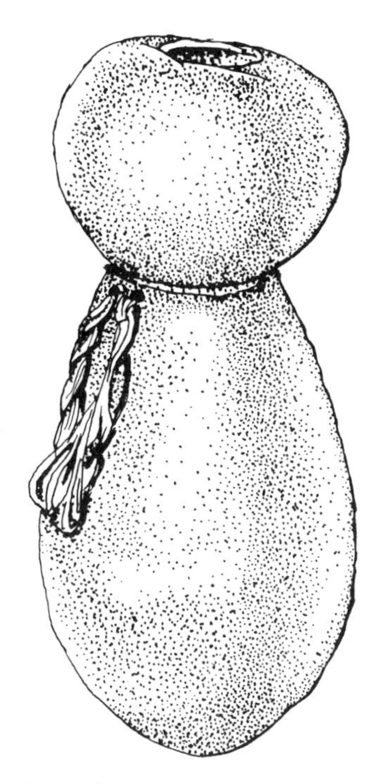

Pā ipu, kuolo, or *ipu pa'i*

rim of the upper gourd for a short distance. Another opening was cut in the natural bottom of the upper gourd to increase the resonance of the drum.

The gourds were permanently joined in one of two ways, a direct and an indirect method. In the direct one the angular space between the outer rim of the lower gourd and the inner surface of the rim of the upper gourd was filled with the gum derived from the milky sap of breadfruit. In the indirect method a circular collar, from another gourd, was fitted over the lower gourd to form an angular recess with the projecting rim of this gourd. This recess was filled with breadfruit gum as in the direct method, thus fixing the collar firmly to the lower gourd. Because the lower opening of the upper gourd fitted over the collar there was a second angular recess; this was also filled with gum, which completed the drum.

Sometimes a cord or strip of tapa was tied around the instrument where the two gourds were joined, with a loop added for fingers. The loop was used to carry the drum and to manipulate this heavy and bulky instrument with one hand while performing. The double drum was sometimes decorated with burnt-in triangles, figures, or dots, arranged in rows.

Two contrasting sounds were obtained with this drum: a deep, dull tone when the player, holding the drum with both hands, dropped it on the hard bare ground or on a pad placed in front of him; and a lighter sharp sound produced when it was held in the air and spatted (slapped) with the right hand. The *kuolo* accompanied the *hula kuolo* and designated chants. Today, a single drum is frequently substituted for this double drum.

The double-gourd drum is found nowhere else in Polynesia.

The Hawaiians had two membrane drums, one of wood, *pahu hula,* and the other a coconut-shell knee drum, *pūniu.*

Pahu Hula (Wooden Hula Drum)

This drum was usually made from a section from the base of a coconut trunk. The bark was removed and the sides slightly trimmed; this exposed the prominent vascular bundles cut at various angles and created an especially beautiful exterior. The top of the section was dug out for about two-thirds of its height to create the resonance chamber. The section was then turned over and hollowed out; this made the base of the drum. A thick septum (diaphragm), convex on its lower surface, remained between the hollowed portions. Occasionally, a trunk of a breadfruit tree was used to make this drum.

A drumhead, a piece of freshly prepared shark belly skin, was fitted over the upper opening, the rim of which had been rounded off so that the skin laid smoothly over it. Regularly spaced holes in one or two rows were made in the ovelap of the drumhead to accommodate the tautening cords that held

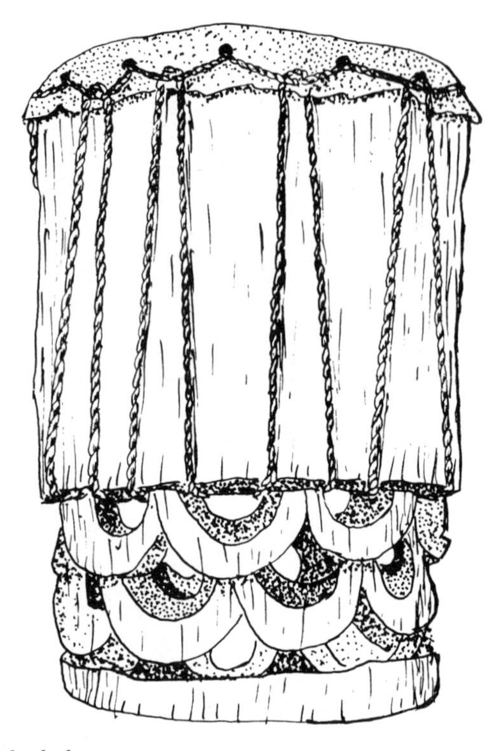

Pahu hula

it in place. Fastened to the drumhead, the cords passed down the sides of the cylinder and, in turn, were fastened to short rectangular studs or "feet" at the base of the drum. In the simplest, and probably earliest, model, these feet were formed by cutting rectangular pieces from the lower rim. A more common technique was to make longer studs, starting these just below the septum, leaving a continuous solid ring at the lower limits for reenforcement of the base to prevent breakage. In some of these drums, the bases were elaborately carved in a single or double zigzag pattern, or in a series of arches in one or two rows, creating a fenestrated design. In an elaborately carved drum the arches were carved as human figures.

A continuous two-ply or sometimes a five-ply sennit braid was used for the tautening cords. They were attached to the drumhead in one of two ways. In the direct method the braid was tied to one of the basal studs, passed upward and directly above, through one of a row of holes in the drumhead, brought down obliquely to and around the next stud, and brought up to the next overlap hole, moving right until completion. In the indirect method, *olonā* cordage was threaded through rows of holes in the drumhead overlap to make different patterns. The sennit braid was passed under and over the threaded *olonā* cordage and in either direct or crossed loops around the studs.

The drum was beaten with the fingertips, the palm of the left hand, or a fist. While involved in the beating of this drum with the left hand, the instrumentalist struck another drum, *pūniu,* with the right hand.

Pūniu (Coconut Knee Drum)

The knee drum was made from the shell of a coconut; hence its name. The top part, the stem end of the nut that contains the "eyes," was cut off level above its middle or greatest diameter, and the outer surface smoothed and polished. The opening was covered with a circular piece of properly prepared skin of a surgeonfish, *kala,* because it has such small scales. The skin piece was large enough to provide an overlap in which one or two rows of evenly spaced holes were made.

To provide for the lower purchase of the tautening cords, a ring *(pōʻaha)* of some material, either tapa coiled around a central core of the same material, a single round of coarse sennit, or several rounds of finer cordage, was snuggly fitted horizontally around the maximum diameter of the shell.

In a direct method of attaching the tautening cord, one continuous piece, it was knotted at one end, and the unknotted end then threaded through one of the overlap holes. Next, the cord was passed obliquely under and over the ring and carried obliquely to the next skin hole and so on. In an indirect method, as in the wooden drum, an *olonā* cord was threaded through the holes in the drumhead over-

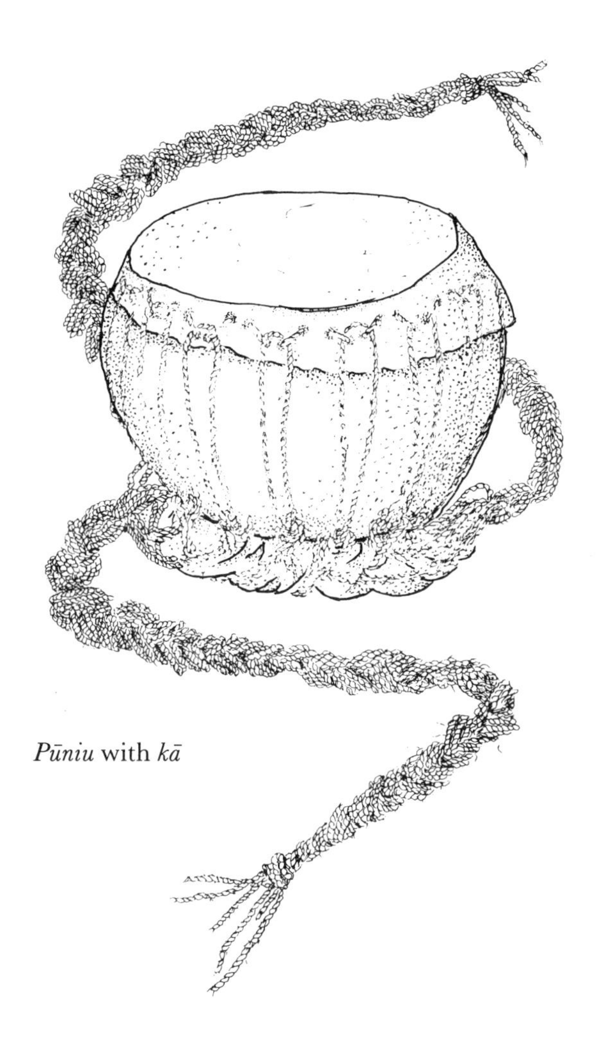

Pūniu with *kā*

lap, and the tautening cord looped over that cord in various patterns.

Two long braided pieces of sennit were attached to the ring to tie this drum to the right thigh of the player, just above the knee.

A beater *(kā)* for this drum was made from a piece of thick two-ply sennit, or dry ti leaves, knotted at one end. Held at the unknotted end, the drumhead was tapped with the knotted end with the right hand. The musician produced deep bass sounds by striking the drumhead of the wooden drum on his left, and accented this with intermittent staccato notes by tapping the knee drum.

The knee drum was used only in Hawai'i.

'ŪKĒKĒ (Musical Bow)

The musical bow was the only stringed instrument in old Hawai'i. The preferred wood for this bow was *kauila,* with *'ūlei* a second choice. Pandanus, sandalwood, and *koai'a* (a small form of *koa*) were also sometimes used.

The end of the instrument to be placed in the mouth was cut away at the back to form a narrow flange. Two or three longitudinal slits were cut or two or three holes drilled in the flange, with the former method the more usual. The distal end of the bow was notched on each side to form a "fish-tail" projection; a median notch was made at its far end.

The strings were made of fine two-ply sennit. The method of stringing the bow with the slit ends was as follows: one end of the string was knotted with an overhand knot and slipped over the slit from the front so that the knot rested in the back recess formed by the flange. In the model with drilled holes, the strings were passed through the holes and knotted at the back. Each string was knotted on the upper side of the bow, drawn taut along the front flat surface to give the bow a slight bend, and then passed around the side notch in the fish-tail projection on its respective side. By making some transverse turns and passing the end of the string under

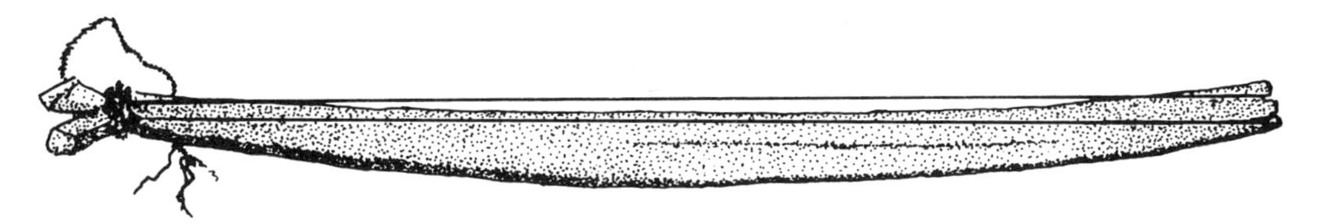

'Ūkēkē

the taut cord without knotting, it was easier to unfasten a string to tauten it when it became slack.

There was very little bend to the bow even when the strings were drawn taut. To gain more clearance the Hawaiians inserted small pieces of gourd under the strings. The bow was played as follows: the instrument was held in the left hand, with the palm pressing the lower edge. The end with the slits was placed between the lips and the instrument extended to the left. The mouth cavity served as a resonance chamber. Using a small fiber or more often a small piece of tapa twisted to form a point, the player rapidly picked the strings to produce a kind of singing or humming.

Although the *'ūkēkē* was used to accompany a *mele* this instrument was also used by lovers to serenade one another.

The wind instruments used by the ancient Hawaiians included the bamboo nose flute (*'ohe hano ihu* or simply *hano*), the gourd whistle (*ipu hōkiokio*), and the Jew's harp (*nī'au kani*).

'OHE HANO IHU OR *HANO*
(Bamboo Nose Flute)

This name is derived from the words *'ohe,* bamboo; *hano,* flute; and *ihu,* nose. It was made from a length of bamboo that included a node and internode. The node was trimmed close to the septum on its outer side with an embouchure (nose hole) made as close as possible to the node on the inner side, the internode. Usually two and sometimes three additional holes were made as finger holes or "stops."

The flute was played as follows: the musician held the instrument horizontally, positioned either to his right, parallel to the front of his face, or pointing outward at right angles to the plane of his face. The nose hole was held close to but not against the right nostril, with the left thumb closing the left nostril by pressing against its side; this increased the draft of air being blown into the nose hole.

'Ohe hano ihu or *hano*

The nose flute was not used as an accompaniment for a *hula,* as most of the other instruments were, but chiefly for lovemaking.

The nose flute is generally found throughout Polynesia except in Samoa and New Zealand, where the bamboo flute is played with the mouth as is usual throughout the world.

HŌKIOKIO, OR *PU'A*
(Gourd Whistle)

This instrument is another that was not used to accompany *mele* and *hula.* It was referred to as a lover's whistle, used by lovers to entertain one another, especially in the evening. They are unique to Hawai'i. Perhaps the great variety in shape and size of gourds grown here contributed to their popularity.

A gourd whistle was made from either a single pear-shaped gourd or a double gourd. The stalk end of the former was cut off to form an end hole, or a hole was made in the side of the pointed end. This was the hole into which air was blown. There were three holes arranged midway down the side of the gourd in a horizontal row, variously distanced from

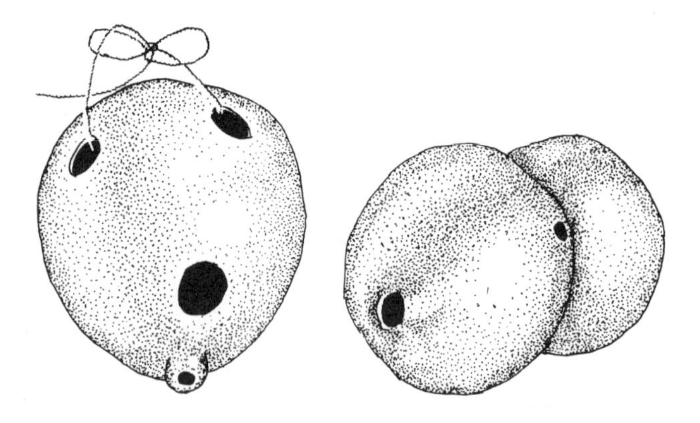

Hōkiokio, hoehoe, or *pu'a:* pear-shaped *(left)* and double *(right)*

the nose hole and from one another. In the double gourd whistle the same types of holes were made, but they were arranged somewhat differently. Some whistles were decorated; the usual motif was small triangles.

Like the bamboo nose flute the gourd whistle was played by blowing air through the nose hole and controlling the sound by stopping the finger holes.

NĪ'AU KANI (Jew's Harp)

This instrument derives its name from *nī'au,* the midrib of the coconut leaflet, and *kani,* to sound.

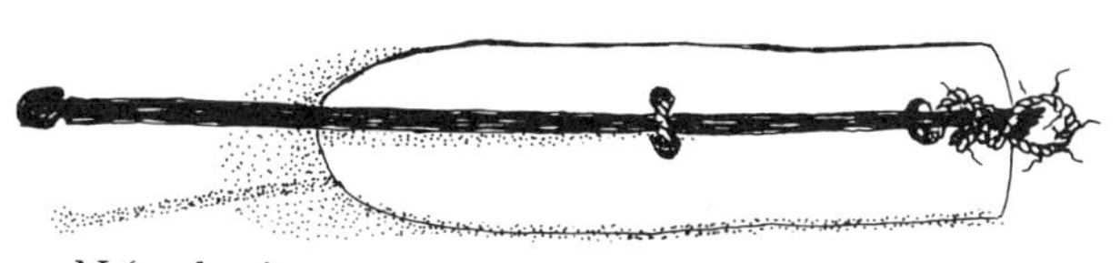

Nī'au kani

The sounding part or vibrator (the reed) was a piece of dried coconut-leaflet midrib *(nī'au)* and was therefore stiff and hard. The instrument consisted of a piece of bamboo or thin wood with a longitudinal slit at one end. The reed was fastened to the bamboo or wood piece with sennit so that it lay flat and firmly over the slit and extended some distance beyond it. The instrument, held in the left hand, was played by pressing one end of the bamboo or wood piece against the lips or teeth and vibrating (flipping) the extended end of the reed with the right hand, at the same time projecting the breath through the slit.

SOURCES

Barrère et al. (1980); Bishop (1940, 52–60); Bryan (1938, 8–15); Buck (1957, 387–416, 456–480); Ellis (1969, 78–79, 100–101, 105, 201); Emerson (1965); Fornander (1918–1919, 208–209), (1919, 594); Kahananui (1960); Malo (1951, 136, 139, 231); Mitchell (1972, 47–60); Roberts (1926); Winne (1965).

8
GAMES AND SPORTS

THE ANCIENT Hawaiians had many games *(pā'ani);* some were played only by children, others only by adults. There were also games played by both children and adults, games that today are played only by children.

MINOR GAMES

For swinging *(lele koali),* the wild morning glory, *koali,* was used in place of the rope we use today. A seat made from a stout branch and tied to the end of a single vine provided a swing for two persons, usually a girl and a boy, with the girl sitting on the boy's legs. A cord or piece of vine tied to the seat was pulled by a third person to keep the swing in motion. Evidently the swingers did not "pump" as we do today. Two other companions pulled cords or pieces of vine tied on either side of the seat to make the ride more exciting.

Two very simple games, the kind that are always so dear to children, in all times, in all lands, were *panapana nī'au* and *moa nahele.* In *panapana nī'au,* children prepared a coconut leaflet midrib *(nī'au)* by stripping away the blade. The game was simply to bend the midrib into a bow and then to release it so it would spring away. Although it was a simple game, children indulged in this play by the hour. *Moa nahele,* or cock fighting, was played with the leafless fern ally, psilotum, known to the Hawaiians as *moa.* Two children sat or stood facing one another, each holding a branched stem of *moa.* These they interlocked and then slowly pulled apart until the branches of one broke. The other child, without broken branches, was the winner and announced his victory by crowing like a rooster *(moa).*

STRING FIGURES

One of the favorite pastimes for adults and children, both male and female, in old Hawai'i was the making of string figures *(hei);* this is similar to our present-day cat's cradle. The simple figures were made by a single person using a single cord, probably of *hau* bast. More elaborate figures were made by two persons using two or more cords. Usually a chant accompanied this pastime, a chant that told the story depicted by the figure(s). There were more than 115 Hawaiian string figures.

TOPS

Tops *(hū)* were made from *kukui* nuts and bamboo slivers, or pieces of the midrib of a coconut leaflet *(nī'au).* Either one or two holes were drilled in a nut to accommodate the sliver or midrib. If the nut had a pronounced "nose," this served as the tip and only one hole was drilled and the sliver or midrib inserted became the stem. If there was no such protuberance, two holes were drilled, one at each end of the nut. The bamboo sliver or piece of midrib was passed through both holes, leaving a short end for the tip, with the other end forming the stem. The top was spun by twirling the stem between the thumb and first finger. This game could be made into a contest; the participant whose top spun the longest was the winner.

STILTS

Both boys and girls, as well as men enjoyed walking on stilts, called *kukuluae'o* or simply *ae'o,* named for the native long-legged stilt (bird) *(Himantopus mexicanus knudseni);* this was also the name of the game itself. Stilts were made from the wood of *'ohe makai.* A length of a main branch became the upright portion and a short section of a secondary horizontal branch the step. The player held the stilts in front of the body and placed the toes and the balls of the feet on the steps that projected toward him or her.

KITES

Flying kites *(ho'olele lupe)* was probably a pastime of both children and adults of both sexes, as it is today. Kites were made of a framework of *hau* wood, which was readily available, easy to bend when fresh, and very light weight when dry; the frame was covered with tapa or plaited *lau hala.* There were a variety of kites: a round one, *lupe lā,* literally a sun kite; a crescent-shaped one, *lupe mahina,* literally a moon kite; a kite with wings, *lupe manu,* literally a bird kite; and a "native" or "genuine" kite called *lupe maoli.* The last is suggestive of European kites and may therefore be of postcontact origin. However, kites of this last-named type are found in many native cultures throughout the world so such a kite may actually have been made and flown by ancient Hawaiians. The framework was made in the form of a cross with the shorter crosspiece nearer one end than the other.

Tails consisted of strips of tapa tied to a length of *olonā* cordage; they were extremely long. Kite cords, also made of *olonā* cordage, might be as long as a mile for large kites; it took two men to keep such large kites aloft. Heavy cords were lashed to stout stakes driven into the ground to secure them. On still days a kite flyer chanted a request to a wind god, perhaps Haluluo'ako'a, to open his calabash and release some breezes. Or he might call to the goddess who controlled winds, La'amaomao, chanting the names of the winds he desired. She would then release these for him. A fifth type of kite was sometimes made, namely *lupe hōkū,* a "star" with six points.

DARTS

Pāhi'uhi'u was a game in which a branch of a tree or shrub was used as a dart. Children simply broke off a branch and removed the bark. The heavier end was sharpened by rubbing it on a rough stone. Adults made a more finished type of dart, cutting off a length of hardwood, trimming it, and sharpening the heavier end. In a special playing field a scorekeeper placed a ti leaf flat on the ground, some distance from the players. Each player held his dart by the slender tip and threw it overhand into the air so that it described an arc and landed with enough force to pierce the ground. The winner was the one whose dart stood upright on or nearest to the ti-leaf goal.

A variation of the foregoing game was *ke'a pua* (*ke'a,* dart, and *pua,* flower) played with a dart made of a sugarcane *(kō)* tassel, the inflorescence. Tassels were removed from the main stalk and laid away to dry. The lower or thicker end of the stem was tightly bound with *hau* cordage and placed in the mouth to wet it. This point was then rubbed in clay to coat it, thus giving weight to the forward end. The game was played by holding the "tail," the dried flowers, with the right forefinger over the end, and the thumb and middle finger on either side of the shaft. The player made a short run forward, and in a stooping position threw the dart underhand so as to graze a mound of earth or sand, and cause the dart to ricochet upward and forward. The player whose dart soared the farthest won one point. The players then went to the other end of the playing field, picked up their darts, and threw them back in the other direction. The player who first scored ten points was the winner.

Although a similar dart game was played in other parts of Polynesia, the Hawaiians were the only ones who used the sugarcane tassel. The playing of this game was limited to the time of the *makahiki,* the period of four months, beginning about the middle of October, reserved for sports and religious festivities; it was also the time when sugarcane flowered.

A variation of the above dart game was called the whip stick and dart game; it had the same name as the other game, *ke'a pua.* As indicated by its common name, the cane-tassel dart was propelled by a stick, usually a rod of bamboo with one end having a lesser diameter than the other. A piece of sennit or *olonā* cordage was tied to the smaller diameter end of the rod, its free end tied in an overhand knot. The sugar-cane dart was prepared as for the previous game. The game was played as follows: the knot in the whip cord was held down with the left thumb, and the rest of the cord was wound in six long spirals around the whip stick, up to its head. With the dart laid on the ground, pointing in the direction of flight, the whip stick was jerked upward and forward; this caused the dart to rise in the air. As the whip cord unwound and the slip knot became unfastened, the dart continued in flight. The winner was the one whose dart flew the farthest.

PEG AND BALL

A peg and ball game, for which no Hawaiian name could be found, was played with a short stick and a short peg, which, sharpened at both ends, was pushed through a hole in one end of the stick so that the peg projected a short distance on each side of the stick. A ball of green leaves, probably a *kinipōpō* plaited from the blades of coconut leaflets, secured to the stick with sennit or *hau* cordage, was tossed into the air by flicking the stick; the ball was caught alternately on the two points of the peg as it fell.

JUGGLING

Kīolaola was a kind of juggling game played primarily by children. Balls *(kinipōpō)* made of plaited green leaves of coconut or *hala* were tossed in the air, as many as five at a time. Players chanted as they juggled. The winner was the one who kept the most balls in the air at any one time.

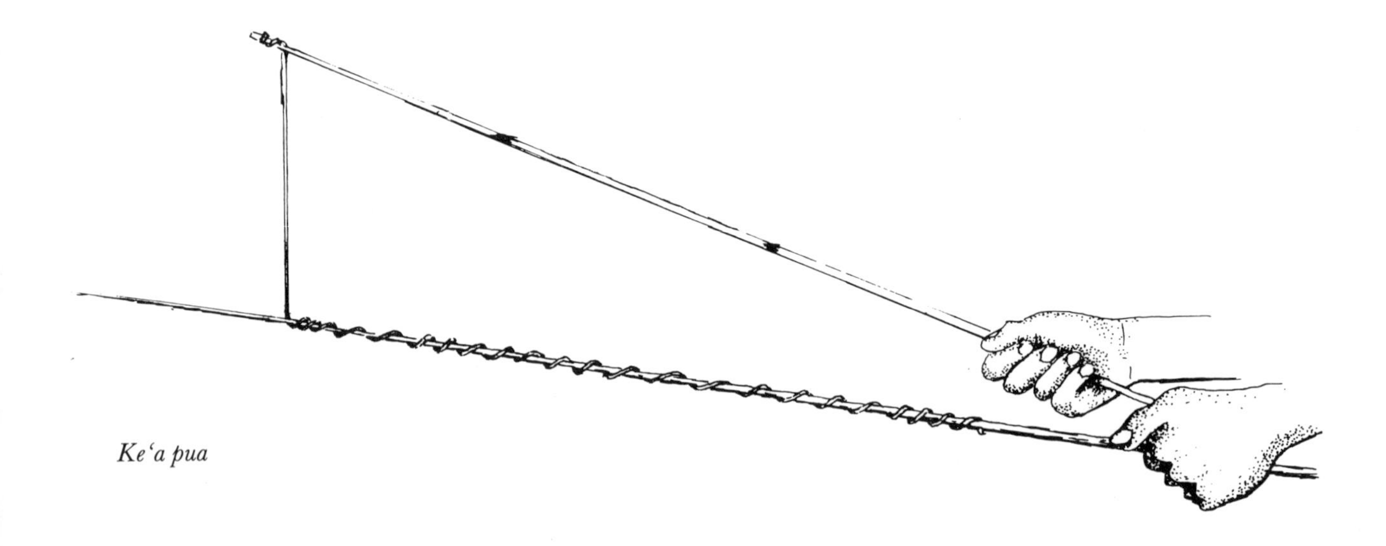

Ke'a pua

BALL AND LOOP

A ball and loop, or ring and ball game as it is sometimes called, has been played in many countries, even today. The old Hawaiian version, played by both adults and children, was called *pala'ie.* The handle was made of either a piece of slender wood or twisted vines (neither identified). The elongated loop was made of two to five pieces of interlaced vine the ends of which were tied to the handle with transverse two-ply sennit. The ball was made of folds of tapa wrapped into an oval form. At one end of this the excess tapa was tightly tied to form the ball. A fine sennit cord was tied to the tufted end of the ball and to the handle about one-third of the distance between its free end and the loop.

The *pala'ie* made today is considered to be a replica of an ancient model but is different in construction from the ones described above. Is the one made today actually similar to an ancient one not recorded? If not, then when and why was the change made since all the materials utilized in the modern version were available in ancient Hawai'i? In the modern model the handle and loop are in one piece, made from braided, cleaned midribs of fresh coconut leaflets, which are flexible and therefore easily braided and bent into a loop. Midribs are braided in multiples of three. The thicker butt ends, where they are attached to the midrib of the frond, are tied together with sennit. Braiding is started from that end and is continued toward the slender tips, where they are bent to form a loop by tying these braided ends with sennit to the handle proper. No set rules for the length of the handle or size of the loop are specified; one uses his or her own judgment or preference. A ball is made from an adequate-sized piece of tapa stuffed with tapa *'ōpala* (small scraps of tapa). The ball is made large enough to sit on top of the loop without going through it. The edges of the piece of tapa are gathered together and tied with a long piece of sennit. The latter is pushed through the ball from the tuft side to the opposite side. To make this operation "authentic Hawaiian," a needle made from the midrib of a coconut leaflet is used. Otherwise, a modern metal sacking needle works well. The farther end of this long cord is then tied to the handle, about one-third of the distance from the butt end to the loop.

Today the objective of this game is to toss the ball upward, within the limits of the length of the cord,

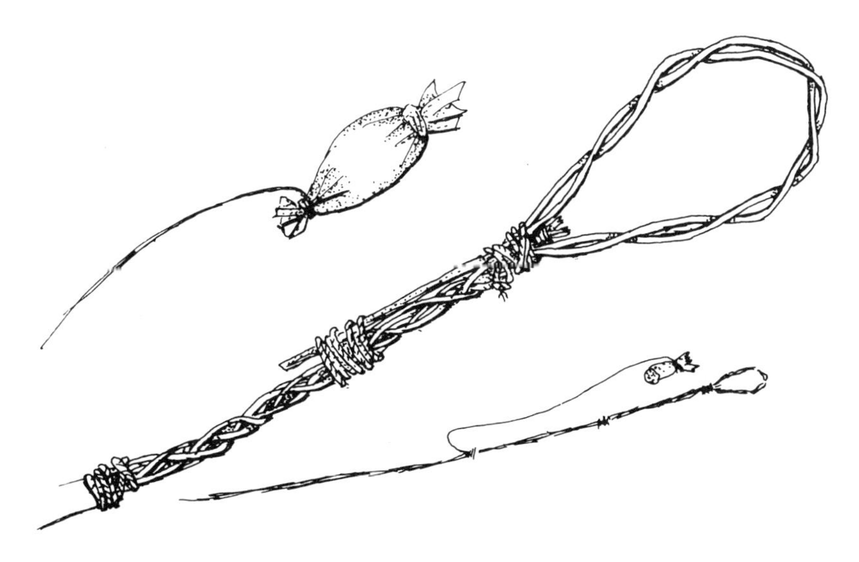

Pala'ie

and to catch it on top of the loop as it descends. In the ancient method the handle was held horizontally and moved in such a manner that the ball described a complete circle as it swung and struck the loop from both above and from below. The complete-circle swinging was repeated to see who could keep making the proper loop contacts without missing. The player making ten consecutive successful circles was winner. Sometimes a group of adults chanted as they played:

> Perch, perch *O'io* bird
> Go up to get us water
> Where is the water?
> In the upland of Nīnole
> On what tree shall I perch?
> On the *'ōhi'a* or the *lama?*
> As the starlight brightens
> And the moon glows
> What light is flickering above?
> It is the Pele family at Kīlauea.
> "In the name of Pele"
> (Beamer, 1987)

CHECKERS

The ancient Hawaiians played a game of checkers called *kōnane.* Both wooden and flat pieces of basaltic rock were used for boards, *papa kōnane.* Only the first is described here. Boards varied in size and number of squares or holes. The "checkers" used consisted of small black beach pebbles and white shells or pieces of white coral, called *'ili 'ele'ele* (black-skinned) and *'ili kea* (white-skinned), respectively. The game was played as follows: the middle of the board was called *piko* (navel) and was marked by an inset human tooth. The checkers were all laid out on the board in various patterns. The game was played by two persons much as Chinese checkers is played today.

GUESSING GAMES

Several guessing games played by groups of adult Hawaiians are reminiscent of games played today. *No'a* was probably the most popular of these. The game was usually played by ten persons of both sexes, divided into two groups of five each, facing one another. On a *lau hala* mat laid between the two lines were placed five bundles of tapa, contiguous to each other. The article to be hidden *(no'a)* was typically a small piece of wood (unidentified), usually in the form of a polished disk with one side flat and the other convex. *No'a* were also made from basalt or coral; these were usually spherical.

Each player held in his or her right hand a highly polished rod or wand made of *kauila* wood. At one end was tied a piece of dry skin with a tuft of shaggy hair or a piece of ti leaf drawn through a hole or slit in the more slender end.

One player on one side was selected to hide the *no'a;* this player took it in the right hand, put the arm as far as the elbow under a tapa bundle, moved the hand about under the bundle, moved on to the next bundle, and somewhere along the line left the *no'a* under one of the bundles. The players in the opposite group had to guess under which bundle the *no'a* had been left; they had watched every movement of the hider's arm and the expression on the hider's face as he or she moved along the line to help them in their guessing. The one hiding the *no'a* for his part had kept his face down and varied his arm movements to prevent this. When a decision had been made under which bundle the *no'a* had been left, the person opposite the suspected bundle leaned forward and gave a sharp flick with a rod to that bundle. This was lifted to show whether the selection was correct or not; if correct that side scored a point. The hiding of the *no'a,* and guessing where it was, alternated between the two groups of players until one side reached the winning total of ten.

The game of *pūhenehene* resembled that of *no'a;* it was played at night, indoors. Instead of the five bundles of tapa, a long piece of tapa made by stitching several pieces together was spread between the two rows of players. The game began by a lead player calling out *"pūhenehene,"* which the whole group repeated. Then one of the men stood up and chanted a happy and pleasing "song." Next, three men lifted the long tapa and covered one of the two groups to conceal it from the other. The covered group then hid the *no'a* on one of its players. When the tapa was removed, the players in the other group tried to guess on whose person the *no'a* had been hidden. The scoring was as in the game of *no'a*. The winning group would dance a triumphant *hula*.

WAND GAME

Commoners of both sexes and lower-rank chiefs and chiefesses played a game called *'ume* or *pili;* it was played in a special house, *hale 'ume,* or in an open court or enclosure. In the latter case the players were seated in a circle around a bonfire. The individual in charge of the game, *anohale,* called the assembly to order. A man called *mau* came forward with a long wand made of highly polished reddish-brown *kauila* wood trimmed with bird feathers. The feathers were red and black, attached to the wand by spiral turns of two-ply *olonā* cordage that passed over the feather quills, which pointed downward. The feathers were arranged near the top of the wand in two circles, slightly separated. Five other sets of circles were evenly spaced over the remaining wand.

The *mau* walked around the circle and touched a man and a woman with his wand. A man desiring some special woman put something of value in the hand of the *mau* and indicated his choice to the latter. Once chosen the couple had to go outside, where they might spend the night together after which each returned to his or her spouse. Evidently there was no jealousy between husband and wife since both participated in this game. The woman could refuse to go through the "engagement" outside; if she did refuse the pair returned immediately to the house.

Kilu was the aristocratic version of *'ume,* with only *ali'i* taking part. The game, similar to quoits, was named after the playing object, a gourd cut obliquely from one end to the other, or an egg-shaped coconut shell cut obliquely from one side to the end with the "eyes." The other object used, the goal, was a conical block of *kauila* wood, broad at the base to keep it upright.

The game was played in the same house or enclosure as for *'ume,* each on alternate nights. Men sat at one end, women at the other, some distance apart. There were five persons per side, each with its own scorekeeper *(helu'ai)*—a man for the men's side, a woman for the women's side. *Lau hala* mats were spread between the two groups. Before beginning to play the scorekeeper called out *"pūheoheo"* and all players answered *"pūheoheo."* Then one scorekeeper, holding a *kilu,* spoke to the other scorekeeper in a low voice: "This *kilu* is a love token; it is a kissing *kilu (kilu honi),*" and then named a person on his side, whereupon the second scorekeeper, also in a low voice, named a player from his side. Next, each scorekeeper gave his *kilu* to the player on his side, each chanting an *oli* before he began the actual playing of the game. A male player slid his *kilu* across the mats with a rotating motion in an attempt to hit the *kilu* of a female player opposite him. If his *kilu* hit the woman's, the scorekeeper chanted an *oli* with a double meaning such as:

> Now wiggles the worm to its goal
> What a tousling; a hasty entrance;
> Pinned; down falls the rain.

The successful player crossed over and claimed a kiss (touching noses on the side) in payment for his success; that was the forfeit the game called for. He followed the kiss with the name of another player sit-

Wand used in game of *'ume*

ting by his side, and his opposite number named one on her side; these two became the next contestants. Sometimes a gift of land or some other possession was substituted for the kiss. Playing continued until someone had scored ten; that person was then declared winner. Such a game might last until morning and sometimes resumed the next night. Often a game of *kilu* was arranged as a compliment to distinguished visitors of rank.

BOWLING

A bowling game, *'ulu maika,* derived its name from the original bowling object used, a disk cut from a half-grown breadfruit, *'ulu.* Later the disks were made of stone. Both the game and the disks, breadfruit or stone, were called *'ulu maika.* The game was played on a prepared course, *kahua;* old maps of the present Honolulu area show many such fields in what is now downtown Honolulu, attesting to the popularity of the game, played primarily by adults. The object of the game was to roll the disk over a considerable distance and between two sticks stuck in the ground a few inches apart.

SIMPLE GAMES

There were also simple games similar to the ones played today where no equipment was required; these included *pe'epe'e,* short for *pe'epe'e akua* (hiding ghost), namely hide-and-seek; *pahipahi,* like our peas-porridge-hot, where the participants slapped both their hands and thighs; and *kuala po'o,* somersaulting, which was often performed in a contest where participants somersaulted down a slope with the first to reach the bottom declared the winner. In *ho'okaka'a,* contestants turned cartwheels along a level course.

MAKAHIKI

Makahiki was the ancient festival beginning about the middle of October and lasting about four months. This was the period when work was suspended and the time to pay taxes with tapa, food, and other products. There were not ordinary but specific religious ceremonies specially honoring Lono, god of farming. There were no wars fought. There were processions in which gods and special banners were carried—and it was the time for many sport events and the playing of games.

Although the celebration of the *makahiki* included many sports, these were not reserved exclusively for this period; young men were engaged in them throughout the year since such sports were part of the training of warriors.

There were the sports played without equipment; actually these were for the most part "training" games, for the preparation of warriors for hand-to-hand fighting. These included *hākōkō,* wrestling while standing; *hākōkō noho,* wrestling while sitting; *uma,* kneeling or prone hand wrestling; *pā uma,* standing wrist wrestling; *kula'i wāwae,* foot pushing; *kulakula'i,* chest pushing; *hukihuki* or *pā'ume'ume,* tug-of-war; *kūkini,* foot racing; *ku'iku'i* or *mokomoko,* rough boxing; *loulou,* pulling hooked fingers or elbows; and *lua* or *ku'i a lua,* dangerous wrestling, including the breaking of bones. These are merely mentioned because no plants are involved.

Then there were the adult sports using darts, clubs, spears, and javelins for throwing and fencing.

JAVELINS (Darts)

Two types of javelins, a long one with one end blunt and the other tapering *(ihe pahe'e* or simply *pahe'e),* and the other shorter and blunter *(moa),* were used in games called, respectively, *pahe'e* and *moa.* Both javelins were made of one of the very hard woods from *'ūlei* or *kauila* trees and were highly polished. The player stood a few yards in back of a starting point, sprang forward, and threw the javelin in such a way that it glanced or slid along the track. The dart was thrown for distance or between two other darts placed a few feet apart. With the first method the winner was the one who threw farthest, and with the second, the one whose dart landed between the two darts without touching either.

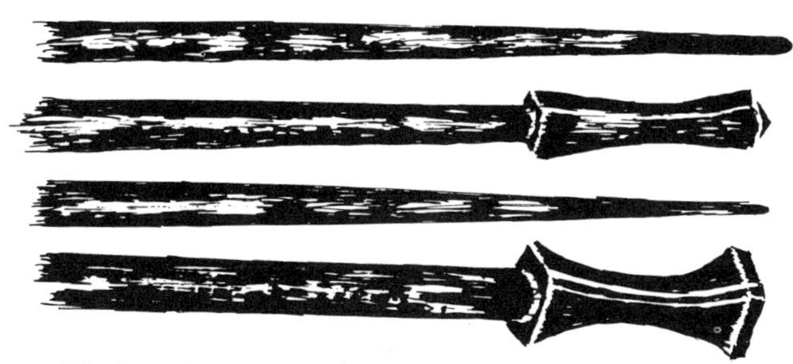
Blade points and butt ends of spears

SPEAR THROWING

In spear throwing (*ʻōʻō ihe* or *lonomakaihe* [named for the god Lono]), the same dart used in the game of *paheʻe* was used as a spear. This sport, as many of the others listed, furnished training of young warriors in preparation for battle. In those training sessions when the point of the *ihe* was exposed, the "target" was either a banana stalk set up for this specific purpose or another warrior. In the latter case the point was padded with tapa, or a lightweight spear made of *hau* wood was substituted for the heavier *ʻūlei* or *kauila* wood spears. There is a story told of Kamehameha the Great, very proficient in this sport, who, when six spears were hurled at him at one time, caught three, parried two, and avoided the sixth by inclining his body slightly.

SPEAR FENCING

Spear fencing *(kākā lāʻau)* was also the name for a short staff or swordlike stick or longer spear made from a hard wood such as *ʻūlei* or *kauila* that was used to strike, thrust, and parry. The object of the sport was to strike or touch any part of the upper portion of the body of the opponent, except the head, with the upper part of the spear, or any part of his lower body with the lower portion of the spear, at the same time warding off the opponent's attempts to strike him.

Kūpololū was a sport in which pole vaulting was done with a long spear *(pololū)*. The vaulting pole was used to pursue an opponent more rapidly than on foot, or, on the other hand, to escape from him. Thus the pole was both an offensive and defensive weapon. The *pololū* was too heavy to throw; instead, when used offensively, it was held firmly at the hip with both hands and the opponent charged.

COCKFIGHTING

Only mention is made of cockfighting, *hakakā-a-moa;* it was much like cockfighting today except that no artificial spurs were used.

BOW AND ARROWS

The use of bows and arrows in Hawaiʻi was limited to the sport of *pana ʻiole,* literally "shoot rat"; it was the sport of kings and chiefs only, played in a specially prepared arena, like a "cockpit." The bows were simple unadorned articles made from a piece of unidentified wood. The two ends were cut down to form shoulders, or grooves were cut near the ends for the attachment of the bow string made of three-ply twisted *olonā* or two-ply *hau* bast cordage. One end of the bow string was permanently tied to one end of the bow with an overhand knot and then a half hitch. The other end was tied in a clove hitch and then a half hitch with a free "tail"; the string could thus be tightened at that end. The arrow *(pana ʻiole)* was made from a stalk of sugarcane, pointed with bone or the hard wood *kauila.*

FOOT RACES

Foot races *(heihei kūkini)* were popular with the ancient Hawaiians, run by both amateurs and professionals, also called *kūkini.* There are wondrous tales of the runs made by *kūkini* that, unfortunately, are beyond the scope of this book.

TOSSING FIREBRANDS

ʻŌahi was a sport carried on during certain times of the year when strong winds swept out to sea from the high cliffs on the Nā-pali coast on Kauaʻi. This unique sport involved tossing firebrands from the cliffs of Makana, Makua, and Kamaile. Participants gathered dried branches of *pāpala* trees on the way to the top of one of these cliffs. There they lighted them and tossed them into the wind. The central core of soft pith of the branches burned rapidly, causing streams of sparks to shoot out like fiery rockets. The light, bouyant branches floated in midair, then rose and fell with the wind. Some of the embers dropped into the sea; others were intercepted by some of the spectators in canoes in the sea below the cliffs and used to brand themselves as a form of tattooing to commemorate the occasion. Such a person was looked upon as a hero. This Hawaiian sport is reminiscent of the proud possession of scars "won" in dueling matches in pre-World War II German universities.

SWIMMING

Living as they did on islands surrounded by the sea, it is not surprising that water sports were a favorite of the ancient Hawaiians. Swimming *('au* or *malo kai)* was not only a sport but a daily function for Hawaiians of all ages and both sexes. It did not include the use of plant materials and so is outside the scope of this book.

BODY SURFING

Body surfing *(kaha nalu)* was a sport or pastime that could be indulged in at any time because no equipment was required. Again, this subject is not treated here.

BOARD SURFING

It was with boards that surfing *(he'e nalu)* reached its highest and unique development. It was an exciting experience to catch a wave properly and maintain one's position, and one requiring great skill.

There were two types of surfboards *(papa he'e nalu):* a short, thin board, usually made of wood from *koa* or breadfruit trees and called *alaia;* and a larger board, in both length and thickness, and cigar-shaped, called *olo.* The front was convexly curved, and the thickness thinned off toward the rounded edges. Although lightweight *wiliwili* was the preferred wood for these larger boards, it was probably difficult to find *wiliwili* trees of sufficient size to make many of them. When such boards were made, they were probably reserved for the nobility, including chiefs.

Olo

The boards were cut out in the rough from trunks of trees of the right size with stone adzes. They were then rubbed down with rough coral to remove the adze marks and finally polished with *'ōahi* (special basalt stones or pumice), as used in finishing canoes. They were then stained with the juice from the ti root, *mole kī,* or the sap, *hili kukui,* from the pounded inner bark of a *kukui* tree trunk. Other stains sometimes used were the juice from banana inflorescence buds, and charcoal obtained by burning *hala* leaves. When the stain was dry the board was rubbed with *kukui*-nut oil. Boards were well cared for: dried after use, oiled, and placed under shelter, and sometimes even wrapped in tapa.

Surfboards were used throughout the Islands along coasts where waves were high and had a long roll before breaking. The short *alaia* boards could, for the most part, be used on waves close to shore, but the long *olo* boards required high waves to prevent them from digging into the surface instead of riding freely on the forward slope of the wave. Surfers using the *olo* board lay on these boards as they paddled with their hands, avoiding breaks in the waves, until they were well beyond the line of breakers. Because waves come in series, it was important to watch for a suitable wave. Both this and subsequent action on the board required great skill and perfect balance. Having chosen a proper wave, the surfer, with his (or her) board directed toward the shore, paddled so as to catch the wave on its front surface; the surfer had to keep paddling vigorously to maintain his position. Having accomplished this, he or she stopped paddling and the wave carried the board and paddler to shore.

The surfer might be content to retain a prone position *(kīpapa),* or, to make the ride even more exciting, by holding on to the sides of the board he could draw his or her body up to a kneeling position, or by placing his or her feet forward he could assume a sitting position. The climax was reached, of course, when he rose from the kneeling position to stand erect.

In surfing it was important to keep forward of the base of the wave while still being carried forward on its front surface. The board could be directed into a slide to the right or left by leaning one's weight to that side, or by trailing one's foot in the water to act like a rudder. This movement to right or left not only added excitement to the ride but could also be used to get away from the wave being ridden in case it began to break behind the board.

Races were held with a "buoy" placed near shore to mark the end of the "run."

Canoe races were conducted much like surfboard races.

SLIDING

Sliding down a hillside on some kind of sled was common throughout Polynesia. The simplest kind of sled in old Hawai'i was the cluster of terminal leaves and the stalk of the ti plant. The sledder sat on the leaf cluster and held the stalk passed forward between his legs as a rudder to steer himself. This ti-leaf sliding was called *ho'ohe'e kī.* A steep, grassy slope made especially slippery after a rain was chosen as the site for this sport, which was primarily a sport of commoners, both younger and older. The *ali'i* engaged in a more sophisticated sledding sport, *hōlua.*

SLEDGES (Sleds)

Hōlua was a sport for young male and female *ali'i.* The term *hōlua* was applied to the sport, to the sledge (sled) itself, and to the sledge course. The sledge was sometimes called *papa hōlua.* It was an unusual sledge compared with Western sleds, being very long and very narrow. A *hōlua* was made of two parts: a pair of runners and a superstructure. The lower edge of each runner was rounded for smoother sliding; the after ends were cut vertically while the fore ends curved upward to prevent these ends from digging into the runway.

The runners were made from the hard native wood *kauila.* They were fastened together (held apart) with crosspieces consisting of an upper bar with a downward leg on either side to fit over the upper edges of the runners. The horizontal bars were made long enough to project slightly beyond the outer edges of the legs with which they formed external angles.

The crosspieces were attached to the runners with

Hōlua

sennit lashing. The cord was passed through holes in the runners and over the crosspieces in several turns.

The upper or superstructure portion of the sledge (the platform) was made of two rails, rounded poles of small diameter and of a length to project slightly beyond the fore and aft crosspieces on the runners. Each was laid on the outer ends of the runners' crosspieces. These rails were kept at a proper distance from one another with pieces of bamboo, attached at intervals by lashing with fine two-ply sennit to the underside of the rails. A space was left free of bamboo crosspieces and fixed with slender pliable bamboo rods to form a semieliptical opening on either side for the sledder to grasp the side rails. The rails were wrapped with spiral turns of white tapa, and a finely plaited *lau hala* matting was spread over the upper surface of the rail frame. The side edges of the matting were carried around the rails, enclosing the ends of the bamboo crosspieces and overlapping the underside.

The completed platform was laid on the crossbars of the runners with the projecting ends of the bamboo crosspieces of the platform fitted against the front of the crossbars of the runners with their projecting notches (angles). The platform was then lashed to the runners with sennit cords tied around the notches in the bamboo crosspieces and corresponding notches in the rail crossbars.

The sledding track or course, *(kahua) hōlua,* was made on the side of a hill by building up rocks in depressions for the foundation; this was covered with earth and beaten hard to create a flat sloping surface. Before it was used, the track was covered with grass. Either *kukui*-nut oil was poured over the grass or the runners of the sledge were "greased" with this oil to facilitate sliding. The track was narrow and could accommodate only one *hōlua* at a time. The track continued from the slope onto the plain below.

A participant grasped the sledge firmly by a handgrip, ran a few yards to the starting place, grasped the other handgrip, and threw himself or herself forward with great strength, falling flat on the sled and sliding down the slope. The handgrips were held and the feet were braced against the last crosspieces on the rear end of the sledge. An expert sledder could slide for 150 to 200 yards down the slope and onto the flat land beyond the slope, sometimes attaining a speed of 100 miles an hour. Some sledders even stood up on their sleds. In a contest the winner was the one who traveled the farthest.

TOYS

As stated in the chapter on musical instruments, several articles sometimes listed as such were actually used as toys or in games.

An *'ūlili* consisted of three gourds pierced, end to end, with a stick. The two outer gourds were stationary while the central one was movable. A string attached to this central gourd was pulled to make a whirling sound.

The *pū lā'i* (ti-leaf whistle) consisted of a half-length of a ti leaf wound in a loose, flat spiral, beginning in a narrow spiral at the tip and ending in a wider spiral at the base. Here it was fastened with a short length of a coconut-leaflet midrib or a splinter of bamboo. It was "played" by blowing into the slightly bent narrower end.

Both a coconut-shell bull-roarer and a *kamani*-nut

whistle were called *oeoe.* Although they were of different sizes and produced different sounds when "played" they were made in the same way: a hole was made through the cleaned shell of the nut, the meat or kernel removed, and the outside of the nut polished. The hole, about 1 to 1-1/4 inches in diameter for the bull-roarer and about 3/8 inch in diameter for the *kamani*-nut whistle, served not only as an opening through which to remove the contents of the nuts but as an "air hole." Two small holes, one on each side of the larger hole, accommodated the ends of a long string (made of sennit for the bull-roarer, and of *hau* cordage for the *kamani*-nut whistle). The nuts were twirled on this string producing, as a result of air trapped in the nuts (entering through the larger holes), a deep sound like a roar, from the bull-roarer, and a high shrill note from the *kamani*-nut whistle.

SOURCES

Bishop (1940, 46–51); Bryan (1938); Buck (1957, 365–386); Ellis (1969, 81–82, 197–200, 209–210, 213–214, 299–300, 369–371); Emory (1965b, 145–157); Finney (1966); Fornander (1918–1919), 146–151, 192–207, 210–217), (1919, 610, 626, 656); Handy and Handy (1972, 351–360); Kamakau (1964, 19–20, 129); Kepelino (1932); Kirch (1985, 271); Malo (1951, 65–68, 148, 155, 214–234); Mitchell (1972, 125–156), (1975).

9
MEDICINE AND MEDICINAL HERBS

The subject of illness *(ma'i)* in ancient Hawai'i—its nature, supposed causes, preparation of the patient, accompanying rituals, treatment, and the practitioners *(kāhuna lā'au lapa'au)*—is a long and involved story.

There were many different kinds of medical practitioners, but this chapter is limited to the *kahuna lā'au lapa'au,* the one who treated with herbs, and the herbs themselves and their uses.

As with comparable beliefs of many native peoples, the ancient Hawaiians believed that most if not all illnesses result from a loss of *mana:* such as the entrance into the body of evil spirits; or from the displeasure of the patient's *'aumakua,* one's personal ancestral god; or from spite, hate, or jealousy on the part of another person; or from one's own "commissions or omissions." If the illness was the result of the entry of an evil spirit into the body of the patient, the *kahuna* whose practice it was to drive out such spirits (the *kahuna kuehu*) was summoned to act first. The *kahuna hāhā,* the palpitating diagnostician, might also be called in to determine the nature of the disease. When these *kāhuna* had performed their duties, the *kahuna lā'au lapa'au* began his treatment.

An interesting aspect of the ancient Hawaiians' concept of dualism was the assignment of masculinity to the right side and femininity to the left side of the human body of both males and females. This concept was carried over into the practice of medicine in this way: it was believed that the right, masculine side was associated with any illness on the outside of the body, and the left, the feminine side, was associated with internal ills. For an example, the practitioner gathered *laukahi kuahiwi* with his right hand to cure a boil, whereas he uprooted the native ginger plant *('awapuhi kuahiwi)* with his left hand for a stomachache.

This *kahuna* was trained through an apprenticeship with an established practitioner beginning as a young boy. It was an honor to be chosen to become such an apprentice; sometimes the son of a chief was selected for this training. The period of apprenticeship was long, ten to fifteen or even twenty years.

One might well ask: why so long? Wouldn't treating an illness with herbs involve only gathering of plants and preparing these in some simple manner, if at all, before administering them? Actually there was much to learn: the names of the plants, where they were to be found, the part(s) of the plant to be used, their preparation, formulations, dosages, and more. This was only part of what needed to be learned; the rest involved learning the rituals and prayers used before, during, and after treatment. As with the study of medicine today, the apprentice needed to learn the anatomy of the body; this was taught through a representation of a body laid out in pebbles on the ground.

Various sources have listed plants used by the ancient Hawaiians for medicine; probably all plants growing in the Islands were used in some way or another for medicine. If one accepts the theory that the value of a plant for medicinal use was determined through the "trial and error" method, the Hawaiians were fortunate that there were no poisonous plants here in precontact times.

What illnesses were most prevalent in old Hawai'i? We are not certain because there are no written records for the time before the arrival of foreigners, and authors who wrote on the subject (after contact) have sometimes included illnesses that were actually introduced by foreigners. Perhaps one way of determining the more prevalent diseases is to note the numbers of plants used for each disease; if these are many, is it too much to assume that such diseases were rather prevalent? As an example, there are many formulations for respiratory diseases. It is not difficult to imagine that there might have been many such diseases in old Hawai'i because the Polynesians who settled here came from islands where the climate was fairly equable the year round; however, here in Hawai'i they found two distinct seasons, one of which was cold and wet. This difference might have been conducive to the development of respiratory diseases.

Another disease prevalent in old Hawai'i was *'ea,* commonly called thrush; this causes lesions or raised white patches to form on the tongue and lining of the mouth. A fungus, *Candida albicans,* is the causative agent. It is primarily a disease of young children; a condition of malnutrition may enable the fungus to flourish in mucous membranes.

Different parts of a plant were used for medicine: roots, stems, leaves, flowers, bark, fruits, and seeds. These were prepared for use by brewing, pounding and extracting the juice or sap, pounding and making an infusion, or the part to be used was chewed and swallowed without any preparation. Plant material was pounded in special stone mortars with stone pestles made for this purpose only. In cases where leaves were used, dosages consisted of a specific number of leaves; specific handfuls of leaves; or the quantity of leaves that, when rolled together, fitted within the circle formed when the tips of the thumb and forefinger were joined. When bark was used, a strip of a designated width and length was prescribed. For berries, flowers, flower buds, and the like specific numbers determined the dosage. The "magic" numbers in prescribing dosages, times and, duration of treatment were one, three, and five; four and five; five and six; or five only, according to different sources. Pounded material was strained through or squeezed out with the cleaned fabriclike sheath at the base of coconut fronds *('a'a niu)* or with the fibers of the native sedge *makaloa.* Medicinal herbs were usually administered in formulations that almost always also included salt and red clay, *'alaea.*

At the end of a period of medication it was the custom for the *kahuna lā'au lapa'au* to administer a laxative, using perhaps the juice of *koali,* or *pi'ikū,* the sap of the green *kukui* fruit. Following this, the patient was given special food, *pani,* to "close" the medical treatment; this was commonly seafood but not necessarily so.

No attempt will be made here to list the names of all plants used for medicine and the diseases for which they were prescribed; only some will be given. The cited examples and their preparation have come to us as folk medicine. Precontact prescriptions have been altered by addition or substitution of postcontact-introduced plants.

Ashes from the leaves of shampoo ginger *('awapuhi kuahiwi)* were mixed with the ashes of bamboo and the sap of young *kukui* fruits to be applied to cuts and skin sores. An infusion of the underground stems of *'awapuhi kuahiwi* mixed with water was used for stomachache, or an infusion was mixed with salt for headaches *('eha ke po'o),* the mixture applied to that portion of the head that was painful.

An extract made by squeezing the sap from macerated flesh from the underground stem of *'ōlena* (turmeric) was used for earaches *(pepeiao 'eha).* In a formulation also containing the rhizomes of shampoo ginger, the extract of *'ōlena* was used for growths in the nose, and *'ōlena,* pounded until soft and baked, was inserted into the nostril for catarrh. A drink made from *'ōlena* and the barks of mountain apple and *koa,* plus various other plants, was used to "clean the blood," an old term used to designate the cure of a variety of illnesses or malaise. A drink made from the underground stems of *'ōlena* alone was used for bleeding ulcers.

The sap of the stem of *pua kala,* the white-flowered prickly poppy, was used to ease the pain of toothaches *(niho hu'i),* ulcers *(pūhā),* and nerve-related diseases.

'Awa was primarily used as a cure for sleeplessness *(hia'ā)* or to induce sleep in case of a fever. For these purposes, and as a tonic *(lā'au ho'oikaika),* in the case of general debility, a drink was made from the roots, and dosages were repeated until the patient became completely relaxed and fell asleep. Although *'awa,* known elsewhere in parts of the Pacific as *kava,* was used as a drink in an elaborate ceremony, in Hawai'i it was used primarily for medicine. However, a coconut-shell cup of *'awa* drink was offered to a high-ranking visitor as a sign of hospitality. Other medicinal uses of *'awa* were for severe headaches; for sore muscles, in a drink made with coconut milk; for lung infections; for difficulty in passing urine; and for displacement of the womb. It was frequently used with other plants in medicinal drinks.

Leaves of *laukahi* were used, either fresh or dried by baking them in ti leaves, as a tonic for both children, especially, and adults. Leaves were also eaten as a laxative. For boils, several leaves were rubbed together with some salt to soften them and applied as a poultice. This was applied as follows: a narrow strip of tapa was twisted and a ring formed with it; this was placed around the boil and the prepared *laukahi* concoction placed within the ring. The whole was then bound in place with a bandage made of a strip of tapa. The mature seeds of this plant have a gelatinous coating that swells in hot water; this drink was used for diverticulitis, as it is today in the Western world under the name of psilla seed.

One of the common uses of *pōhuehue,* the beach morning glory, was for sprains *(māui).* The vine was mashed, placed over the injured limb, and bound in place with a tapa bandage; salt was probably added to this compress. The pounded roots of this vine made an effective cathartic, *lā'au ho'onahā.*

The bark of *'ōhi'a 'ai,* mountain apple, was chewed and swallowed for sore throats. The bark was also mixed with entire plants of the *'awikiwiki* vine and the entire plant of *'ihi,* native purslane species; pounded thoroughly; water added; mixed; strained; and heated. When cool this was applied to itchy skin or kindred skin disorders. A formula containing the bark, leaves, and young buds of the mountain apple tree along with *kukui* flowers and the buds, flowers, and leaves of *'uhaloa* and several other plants was used as a tonic for young and older children as well as adults.

Ti leaves were dipped in cold spring water and placed on the forehead of patients with fever and headache to cool and help relieve pain. For a "dry fever," a fever not accompanied with perspiration, the midrib of the leaf was removed and the remaining halves of the blade tied end to end to make a "belt." This was tied around the chest or abdomen. This treatment was repeated twice a day until the fever had passed.

The slimy sap of *hau* stems and flower buds, and the bases (ovaries) of open flowers were given as a mild laxative to both children and adults; the dosage varied with the age of the patient. Sometimes the ovaries of the related native hibiscuses, *aloalo,* were used instead. Very young shoots or flower buds of *hau* were chewed thoroughly and swallowed for "dry-throat," a condition that sometimes prevails during a cold when phlegm collects in the throat and the latter "tightens up." The slimy sap loosened the phlegm and relaxed the throat. The sap, mixed with water, was given to a woman while delivering a baby to ease the birth process.

The prepared starch from *pia,* Polynesian arrowroot, was mixed with water and given several days in succession for diarrhea. In severe cases, such as in dysentery, the red clay *'alaea* was added.

The ripe *noni* fruit was used as a poultice for boils in the same way as *laukahi.* The flowers and fruit were eaten for kidney and bladder disorders. For deep cuts the juice of a half-ripe fruit was squeezed onto the area. *Noni* was also used as one of the ingredients in various formulations for constipation.

The leaves of *pōpolo,* the glossy nightshade, were used for medicine, and without the addition of any other plants. Young leaves were eaten raw along with a meal to prevent bloating; these were also used, either raw or cooked, for coughs. Also, either raw or cooked or made into a tea they were used as a tonic. For eye problems, such as inflamed eyes, a compress of mashed leaves was applied. Leaves pounded with salt were used for wounds in general.

Both the flowers and the sap from the green fruits of *kukui* were important in the treatment of the children's disease *'ea* (thrush), which causes slightly raised white patches on the tongue and the lining of the mouth. The flowers were either crushed or first chewed by an adult, if the child had no teeth. The fruit stem was broken off at its point of attachment, leaving a little "well," which then fills with sap. The sap was picked up with a fingertip and rubbed on affected areas. Children left weak as a result of stomach and bowel disorders were given a tonic composed of *kukui* flowers, stems of one of the native peperomias, bark of the mountain apple, roasted *kukui*-nut kernel, a mature *noni* fruit, and sugarcane juice. These were ground together, the mixture pressed out, and then strained; the liquid obtained was administered three times a day.

A scrofulous sore or bad case of skin ulcer was first washed with an infusion made by boiling *'ahakea* bark in water. Then a mixture of pulverized wood from *lama,* pounded roasted *kukui*-nut kernels, and the milky sap of the breadfruit was absorbed on *makaloa* fibers to form a compress to apply to the cleaned sore.

A mixture of roasted and crushed *kukui*-nut kernels, roasted young taro leaves, and the very young fronds of the *kikawaiō* fern was eaten as a tonic by a patient recovering from a recent serious illness.

The leaves of the following plants, *wauke, mamaki, ko'oko'olau,* and many more were used to prepare hot teas as tonics and general "cleansing agents." Leaves were used both fresh and dried.

The *kahuna lapa'au lā'au* knew how to set broken bones and to treat the resulting swelling with the *koali* vine; if the fracture was a compound one he prevented infection by applying a mixture of crushed *koa* bark and *noni* leaves.

As with all native peoples, the Hawaiians also had faith healers of the type referred to today: one who heals by prayer and/or ritual alone without the aid of medicinal preparations and/or physical therapy. These are not discussed in this book.

Physical therapy is noted here but not reviewed in any detail. This type of treatment included massage *(lomilomi),* without the use of oils; the steam bath *(pūholo),* involving steam created by pouring cold water on hot stones, combined with leaves of undesignated plants; sweating *(hou),* wrapping the patient in ti or *'ape* leaves, or applying hot dry packs made of heated salt wrapped in tapa; immersion in fresh or salt water; "bleeding," bloodletting *('ō'ō);* and the use of an enema *(hahano)* made of warmed seawater with an infusion of the sap from the wild *'ilima* or from the green *kukui* fruit. These enemas were administered with a syringe *(hano)* made from a joint of bamboo or a long-necked gourd.

There are no longer *kāhuna lā'au lapa'au* of the type found in old Hawai'i because there have been no kahunas of this kind to teach apprentices since the early part of this century, but their knowledge and practices have in part carried over to today as folk medicine. There are Hawaiians today who know many plants of medicinal value and their uses without having been trained in the old manner. Also, there are naturopaths who are permitted to practice with the use of native and introduced plants.

SOURCES

Fornander (1919, 608–611); Gutmanis (1977); Handy (1940, 201–205); Handy and Pukui (1972, 142–145); Handy et al. (1934); Kamakau (1964, 94–115); Larsen (1965); McBride (1972), (1975); Malo (1951, 107–111); Mitchell (1972, 193–219); Wise (1965c).

10
WAR AND WEAPONS

Because one thinks of the ancient Hawaiians as working in the *lo'i* and tending other crops, fishing, going into the forest for trees to build their houses and canoes, gathering herbs for medicine, and collecting feathers of brightly colored birds for their capes and helmets (all involving hard work but peaceful in intent), it comes as a surprise to learn how warlike these people were, and how many battles they fought.

The causes of their wars were personal vengeance for loss of and insult to prestige, revenge for ill treatment of an honored leader, or the result of boredom, but the most common cause was dispute over land.

Each king and high chief maintained a nucleus of an army as part of his court. These men formed a special group, the warriors, who were trained in methods of hand-to-hand combat and the use of weapons as part of their sports. However, any able-bodied men could be conscripted for battle service.

As part of their training there were sham fights *(kaua kio)*, usually with blunted spears; if sharp spears were used such an engagement was called *kaua pahukala* or simply *pahukala*. Many chiefs were killed during such mock battles.

The chief counselor to the king *(kālaimoku)*, an office similar to that of a prime minister in Europe today, was well-versed in the principles of warfare: how to set a battle in order, how to conduct it correctly, and how to adapt the battle to the ground on which it was to be fought.

The seer *(kilo)* had to be sure that a battle would be favorable to the gods; among these might be Kū-ho'one-nu'u, the war god of Maui, or if it was Kamehameha I going into battle, it would be the feathered god Kū-kā'ili-moku. If the omens were good the king and/or chiefs fixed the time and manner of mobilization. Messengers *('elele)* were sent to all districts under the king's or chief's command to inform lesser chiefs of the coming battle and how many men were needed—not only warriors but any other able-bodied men.

Before a battle was actually fought, the *kilo* once more consulted the gods, studied omens, and made appropriate sacrifices, usually animals but in the case of a very important battle, humans. Just before the battle began the king sent for an astrologer *(kilo lani)* to get his opinion about the coming battle. The *kilo lani* then determined by studying the heavens whether the time was favorable *(kūpono)* or not *('āpuni)*.

There was a regular order to the disposition of groups in the assembled army. A priest *(kahuna)*, who could be compared to a present-day army chaplain, went ahead of the main body, bearing a *hau* branch. The priest set this upright in the ground and guarded it as a favorable omen or sign for his side. Each side respected this emblem of the enemy and did not interfere with each other's *hau* symbol, their *mīhau*. As long as the *mīhau* stood erect it meant victory for that side, but the losing side would let its *mīhau* fall. While this *kahuna* served on the battlefield, a large body of priests was praying and offering sacrifices for victory in the war temple *(heiau waikaua)*.

Behind the *kahuna* carrying the *hau* branch were the groups of warriors and "enlisted" men. Usually the king, his wife, his war god(s) carried by the *kahuna ki'i,* and closest friends were in one of the rear groups, where these important persons were somewhat protected by other groups. At the rear came those wives who did not have to remain at home to care for young children; they carried food and water for the army and tended the wounded left behind. How reminiscent this is of stories in older textbooks of women following the armies in Europe in earlier days.

As the two armies approached but were still some distance from one another, men with slings exchanged volleys of stones; as they approached closer, but still at some distance, javelins and spears were thrown. Then, closing in for hand-to-hand fighting, they began using daggers and clubs.

The first man killed was called *lehua.* Could the reference be to their valued *'ōhi'a lehua* tree—the spilled blood of the fallen warrior likened to the red flower of this tree? If the body was that of the enemy and could be obtained by the victors, it became a choice object for sacrifice.

Battles were bloody affairs; accounts indicate that a whole army might be annihilated. When fleeing after losing a battle, the remnants of an army were pursued by the victors and "cut down." Victors buried their dead, but the bodies of the losing army were left to rot or be eaten by animals. War was as terrible and ignoble in those days as it is today.

If both sides were evenly matched, or if for some other reason a truce was called, a treaty of peace was solemnized over the blood of a slain pig, followed by feasting and sports—the very sports in which the arts of war had been learned.

Although most battles were fought during daylight, some night fighting did take place, when a surprise attack, an ambush *(moemoe),* might be made.

This is a brief account of ancient Hawaiian warfare—all that is warranted in a book on ethnobotany but necessary as a background for a description of the weapons used.

Among these weapons were various types of javelins and spears, *ihe* and *pololū,* respectively; many different kinds of daggers *(pāhoa)*—daggers were peculiar to Hawai'i in Polynesia; a great variety of clubs *(lā'au pālau);* and slings *(ma'a).*

SPEARS

There were two kinds of spears: a long spear *(pololū),* which was used like a pike; and a shorter one *(ihe),* used as a javelin. All spears were made from a single piece of wood, *kauila* being the preferred wood. *Kauila* is a hard dark brown or dark red wood that takes a beautiful polish; it has the appearance of mahogany.

Not only were the long spears distinguished by their length but by the enlargement at their butt end. The spear shaft was solid, and circular in cross section. The point was flattened to form lateral edges; this extended back for various distances from the actual point and is therefore classified as a blade point.

The blade point was formed by trimming the end of the round shaft into two surfaces. Each surface consisted of two planes that sloped forward from a median edge. The lateral planes from opposite surfaces met to form lateral cutting edges. The actual point was formed by curving the side edges into a median point, thus giving the blade point a lanceolate form. In cross section this was lozenge-shaped, with the greater diameter, from side to side, being the width, and the lesser diameter, between the median edges, the thickness. Sometimes the blade point was scraped so that there was no median edge; these blades were thinner. The actual blade point was spatulate in form.

In the preliminary shaping of the wood the butt end was left larger than the adjacent shaft, then trimmed down to the desired form. The shape of the butt enlargement varied; the most common form was rectangular with near and far ends of this four-sided enlargement of the same thickness but with the surfaces between concave. The resulting middle constricted portion remained four-sided, but the edges were usually trimmed down to form a smoother grip. Variations of this rectangular enlargement included cutting down the far or near end at a slope, or cutting either of these ends at right angles. In another type of enlargement the butt end was circular in cross section instead of rectangular; in a double form the butt end was in two parts, one circular and the other rectangular in cross section. In still another type the butt enlargements ended in a proximal point. In this group the enlargements were circular or rectangular in cross section, or circular in cross section with a rounded knob covering the point. As in the second type just described there were also double forms in this third type, with two circular parts, or one circular and the other rectangular.

JAVELINS

The points of the short spears or javelins *(ihe)* were either barbed or unbarbed. The barbed javelins *(ihe laumeki* or *ihe laumake)* consisted of a shaft tapering to a blunt end at the butt end and gradually increasing in thickness within a few inches of the point, the latter tapered suddenly into the barbed portion. The barbed portion itself was usually defined by a distinct shoulder, with the shaft cut down sharply. There were four to six rows of barbs. Each barb had a somewhat lanceolate point with two upper surfaces sloping out from a median edge. The barbs in each row alternated in position with the barbs in the row above so that the base of each barb originated from the part below and between the two barbs above. All the barbs sloped outward and downward, toward the butt end. This alternate arrangement of each row made every other row appear as if it was sticking out more than the one immediately next to it.

There were two kinds of unbarbed javelins: those with a blade point and those with a round point. Those with a blade point resembled the shorter forms of the *pololū* spears except that they did not have the butt enlargements of the latter; the end of the shaft was cut off square. Most of the blade points were lanceolate or acutely lanceolate with median edges; also, the shafts of these javelins were less massive than those of the long spears.

In the unbarbed javelins with a rounded point the latter were evenly trimmed around the whole circumference of the shaft at the shaft junction so that the cross section at that point was circular.

There was a type of two-in-one device, a combination pole and spear that had a blade point, which permitted its use as a weapon, but also had hooked and notched projections at either end so that items could be carried without slipping off when it was used as a carrying pole.

DAGGERS

The general Hawaiian name for a dagger was *pāhoa*. The Hawaiians were not only unique among Polynesians in having daggers, but also they had at least five distinct types: truncheon, bludgeon, long-bladed, curve-bladed, and shark-toothed.

The truncheon dagger, made of *kauila* wood, can be characterized as resembling the cut-off pointed end of an unbarbed javelin to which a wrist loop was attached. The butt end was cut off square and was circular in cross section. It was gradually thinned down toward the pointed end to form a blade, with side edges but no median angle(s). The two ends of the wrist loop, made either of *olonā* cordage or plaited five-ply sennit braid, were passed through a rectangular hole made at the center or toward either end of the shaft, and knotted. This is the first mention of a wrist cord on a weapon; it is the method of attaching all wrist cords on weapons in the following descriptions. The pointed end of the truncheon dagger could be used for stabbing, and the blunt end for striking, like a present-day policeman's truncheon.

The bludgeon dagger, as the name indicates, was a combination dagger/club; the butt end of this weapon was enlarged to give weight and thus form a club. Bludgeon daggers were shorter than truncheon daggers. The perforation for the wrist cord or loop was made halfway between the two ends, thus facilitating the use of this weapon as either a dagger or a club. In some of these weapons the club portion was a smooth rounded head with a diameter not much greater than the shaft; others had a large pronounced round head. Still other bludgeon daggers had irregular massive heads, cut from a root of a tree. Some had somewhat flattened blades with convex or rounded points. The shaft was usually circular in cross section.

Long-bladed daggers were characterized by a long double-edged blade ending in a sharp point. The butt end was shaped into a handle, narrower than the blade and meeting the latter on each side in a sloping shoulder. The blades differed in length and shape. The shape varied from one that was narrow, not much wider than the diameter of the handle, and had median edges, to one that had a long double-edge blade, wide at the shoulder where the blade met the handle and tapered to a fine point, with median edges extending the full length of the blade. A wrist loop might or might not be attached; if present it was placed near the butt end of the handle.

The curve-bladed daggers consisted of three sections: handle or grip, shaft, and blade. At the butt end of the handle all surfaces were trimmed down to form a rounded grip with a terminal knob near the end. A perforation for a wrist loop was made where the handle met the shaft. The rectangular cross section of the shaft was greatest at the handle end, diminishing toward the blade end. The latter started with a concave curve on the upper surface that was cut down deeply to form a flat surface; there was a complementary convex curve on the lower surface, which was rounded. The sides of the blade converged to form a sharp point.

The shark-toothed dagger was primarily a club, being a hybrid between a shark-tooth club and a bludgeon dagger. It had a flattened expanded head, like a paddle, with shark teeth attached to its periphery, the number depending upon the size of the head. The handle consisted of a shaft pointed at the end. There was a wrist loop attached toward the pointed end or in the middle of the handle. The fixation of the shark teeth is described later in connection with other shark-teeth weapons.

The Hawaiians also made daggers from swordfish blades (bones).

CLUBS

The short clubs had expanded heads and short handles cut off square at the butt end. Some were supplied with wrist loops, others not. Where a loop was present, the perforation for it was near the butt end. These short clubs are classified according to the form of the head: smooth-, ridged-, rough-, and stone-headed.

The smooth-headed clubs had smoothly rounded heads that sloped gradually to the handle without a

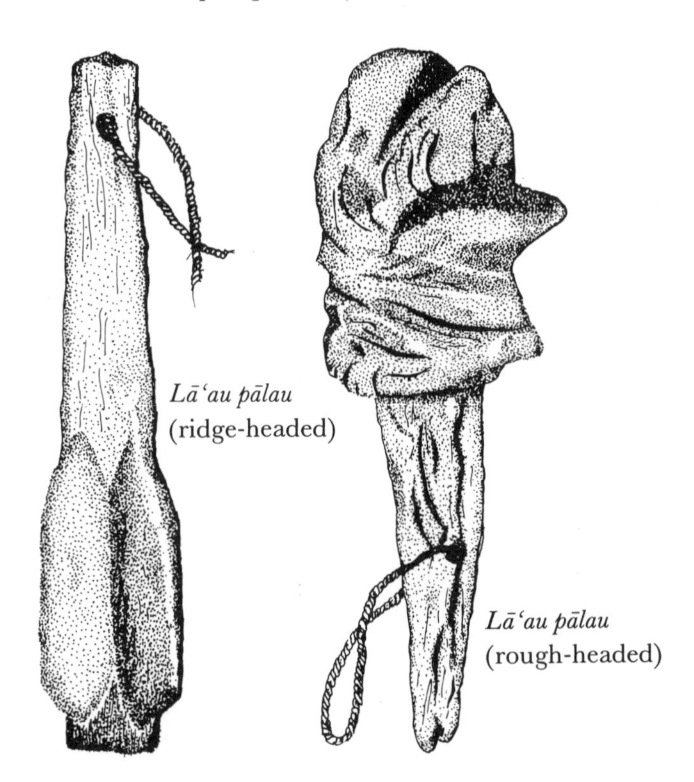

Lāʻau pālau (ridge-headed)

Lāʻau pālau (rough-headed)

shoulder. The butt end was slightly flared. Ridge-headed clubs were just that; instead of a smooth, rounded head, the head was ridged.

Rough-headed clubs had gnarled heads made from natural root or limb enlargements. The heads were only slightly trimmed, but the handles were properly trimmed. The butt end of the handle was rounded, and if a wrist loop was present the perforation for it was near the butt end.

The subject of stone-headed clubs is not within the scope of this book except for a description of the wooden handles and the plaited sennit used to bind the stone heads to the wooden handles. The stone heads were of various shapes, showed beautiful workmanship, and were rather handsome; all of them had two, four, or six longitudinal grooves with uniform oviform edges into which were fitted the cordage used to attach the stone head to the wooden handle. The handle was made from the dark hard wood *kauila,* as were most daggers and clubs.

The handle of the stone-headed club was round in cross section, and was fitted to the lower end of the stone head as follows: the sennit turns (lashing) were passed through transverse holes in the wooden handle, near the junction of the head and handle, and along the bottom of the grooves in the head, to cross over the far end, the top of the head. Three sets of transverse turns were spaced over these longitudinal turns. The middle set of these was above the projecting rim of the lower end of the head, toward the handle, to prevent the sennit turns from slipping down. A wrist loop was attached at the lower end of the handle.

There were four types of shark-tooth clubs: one with a solid elliptical frame and straight handle; one with an open elliptical frame with a straight handle; one with an open elliptical frame and a ring handle; and one with a curved bar with a straight connecting handle, the teeth attached to a cord. All of these consisted of three parts: teeth, handle, and fixation elements. The shark teeth used were of two forms: regular triangular, and triangular with basal cusps. The teeth came from the species of sharks called *niuhi,* large gray man-eating sharks in the genus of *Carcharadon.* The teeth for all shark-tooth clubs were prepared for mounting or fixing by separating them individually from each other, with the cartilaginous base belonging to each tooth in which they were embedded kept in place. One to three small holes were bored neatly through the enamel of the tooth at the junction with the cartilaginous base. This left enough of the base below the holes to be set into slots in the wooden handle.

The handles were usually made of *kauila* wood; the slots were cut so that the bases of the teeth fitted exactly into them. In some weapons of the shark-tooth type the teeth were attached to the side of the handle; in others, at the end of the handle. Small holes of the same size and number as in the teeth were drilled transversely across the slots in the wooden handle, or below the bottoms of those slots. Fine *olonā* cordage or thread, and small slender wooden pegs were used to fix the teeth to the handle. In some cases gum, perhaps from breadfruit, was used to set the teeth, or small wooden wedges were inserted to tighten the teeth.

The teeth were fixed (fastened) in one of five ways: (1) lashing, the most frequently used method, where a two-ply thread of *olonā* was laced between two sets of holes, one in the tooth, the other through the handle below the slot, to form a simple V pattern, an M pattern, or a continuous pattern; (2) direct pegging, where two holes, usually, were drilled transversely across the handle slot and in the base of a tooth, with a peg of hard wood then passed through the holes on either side of the slot and through the tooth; (3) lashing to a peg, where a hole was drilled across the handle slot below the tooth base, with a peg driven across the slot and a lashing thread passed in turn through three holes in the tooth and around a peg; (4) anchoring to a peg, where a hole was drilled at each side of the base of the tooth and the tooth set into the end of the handle, with the *olanā* thread being passed through the holes in the tooth and then through twin tunnels on each side of the wooden handle or shaft; and (5) sewing, which was similar to direct pegging except that, instead of a peg being passed through the holes in the teeth, two pieces of *olonā* string were used, one on each side, to cross at each hole. One or more of these five methods were used in the weapons described in the following paragraphs.

The simplest of the shark-tooth devices was made either of a short wooden handle with a shark tooth attached in the middle, called a *leiomano,* or with a long wooden handle with a tooth attached on one end.

Then there were the devices, also with a single tooth and wooden handles, that were not straight, as were the two just described, but were curved or obtuse-angled. These handles were carved from *kauila* wood, and each had hand loops. An unusual type had a lozenge-shaped handle with knobs at three angles, and a tooth at the fourth angle.

Devices with two teeth included a crescent-shaped handle with a tooth attached to each end, and one that had a crescent-shaped handle with a second limb with a lesser curve to make a kind of double crescent. Again, a shark tooth was attached to each end. This device resembled the knuckle-duster fairly common in the Western world today. Finally, there was the device with a straight handle and a tooth attached to each end.

A shark-tooth club with a solid frame and straight handle was evidently quite common. The frame varied in size and shape although the latter was usually elliptical. However, some of the larger ones had parallel sides. The sides and end of the frame were grooved to accommodate the bases of teeth. The number of teeth varied and both types of teeth were used. The teeth were usually attached with pegs, with wooden wedges driven into the spaces between the tooth and sides of the slot for greater security. These weapons were bisymmetrical, having an even number of teeth on each side, with an occasional odd one at the far end. The curved points and cusps of the teeth were always directed toward the handle because the weapon was used with a draw stroke toward the person with the club, thus making the curved points more effective. Wrist loops were provided.

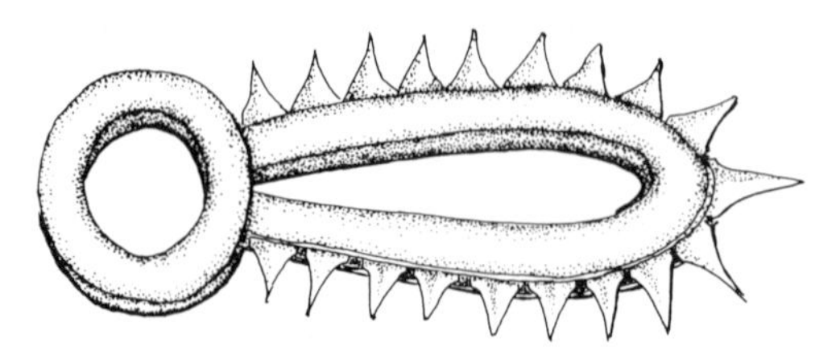

Shark-tooth weapon

The open shark-tooth club with an open frame and straight handle resembled the one with a solid frame (just described) except that the center, the interior, was removed.

The shark-tooth club with an open frame and a ring handle resembled the preceding except for the shape of the handle. This club was carved from one single piece of *kauila* wood.

The shark-tooth club with a curved bar and connecting handle had the same appearance as the crescent-shaped, double-bar club with two teeth except that instead of two teeth, one at each end, it had a total of up to nineteen teeth arranged on the outside of the outer bar. Pegging, direct lashing, and lashing to a peg were all methods used to attach teeth in this weapon. It was also called a knuckle-duster by foreigners who first saw it.

The last addition to the list of shark-tooth weapons was a device consisting of three or more cusped teeth attached in a row to a piece of cord made from five-ply sennit braid or three-ply hibiscus braid, one end of which was doubled back to form a small loop and a double part to hold the teeth. The teeth were set between the two cords and embedded tapa or hibiscus fiber for rigidity. The bases of the teeth were ground vertically so that they fit closely together. A lashing thread of two-ply *olonā* fibers was used for lashing.

TRIPPING WEAPON

The *pīkoi* was a wooden or stone weight, with or without a handle, to which a long cord was attached; this was used as a tripping weapon. After a warrior tripped an enemy, he "finished the job" with a dagger. Tripping weapons without handles were made in various shapes: oval and circular in cross section, with the smaller end coming to a blunt point; shaped somewhat like the foregoing weapon but with the broad end grooved slightly to form a three-lobed triangular base; and with a deep groove across the base to form two rounded knobs. Holes for the throwing cord were in the narrower ends of all weapons. The cord was usually made from *olonā.*

The wooden tripping weapon was made not only of *kauila,* as were the other tripping weapons listed, but also from the wood of *pua.* Some of the heads were made from a smoothed root knot; others had well-carved spherical heads. One was even made with a large flared head. Eight-ply *olonā* braid was used for the cord. The hole for the cord was bored near the butt of the handle.

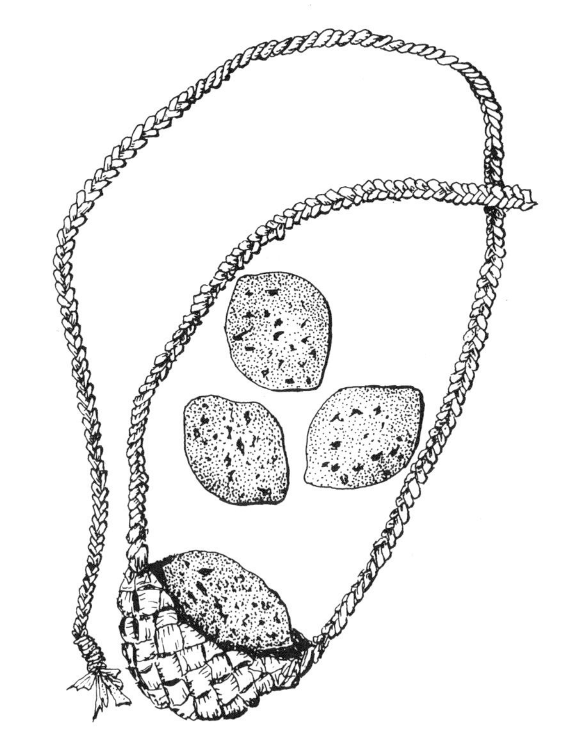

Ma'a with *'alā o ka ma'a*

SLINGS

Slings *(ma'a)* were used throughout Polynesia; Hawai'i was no exception. The pouch was plaited strips of *hau* bast. In one type of sling the pouch narrowed toward the ends by combining two longitudinal strips into single wefts. When this elliptical-shaped pouch reached a length of several inches, the strips at the ends were divided into three plies, which were then braided for the strings.

In another type of sling, instead of a pouch two pieces of sennit braid were used; these were kept separate and the plies of the two braids then combined into one braid for a short distance. This was then changed to three-ply cords to form the strings. The two short braids in the middle were separated to form an open pouch on which the sling stone was placed with its long axis across the length of the open pouch.

Sling stones *('alā o ka ma'a)* were usually made from basalt and highly polished.

SOURCES

Bishop (1940, 39–40); Buck (1957, 417–465); Ellis (1969, 140–162), Emory (1965c), Kirch (1985, 198); Malo (1951, 196–197, 203–204).

11
RELIGION

Religion was part of every phase of life of ancient Hawaiians. They had many gods to whom they called for help in their individual endeavors and guidance in their activities. *Ali'i,* warriors, craftsmen, and commoners all had their particular gods.

GODS

Polynesians brought with them their gods Tu, Tane, Rongo, and Tangaria who became Kū, Kāne, Lono, and Kanaloa. Kū, meaning upright, represented male generating power while Hina, prostrate, was the expression of female fecundity and the power of growth. Kū also referred to the rising sun and Hina to the setting sun. Kū was best known as a god of war. He sometimes assumed the form of the *'ōhi'a lehua* tree, or *'io,* the Hawaiian hawk.

Craftsmen worshipped some form of Kū, one of the several gods of the forest: Kū'ālanawao was the god of canoe makers; Kūhuluhulumanu, the god of fowlers and feather craftsmen; Kūka'ō'ō, the god of the digging stick and therefore of farmers; and so forth. However some professions had gods outside the family of Kū, such as the medicinal practitioners who had Ma'iola; hula dancers, Laka; makers of tapa, Lauhuki; and those who decorated it, La'ahana.

Kāne, also the Hawaiian word for man, was the leading god among the four great gods of ancient Hawai'i. He was the god of all creation and the ancestor of chiefs and commoners, a god of sunlight, fresh water, and forests. As such, various forms of Kāne were adopted in connection with natural phenomena: Kānewahilani, god of the heavens; Kānelu, god of the earth. Although he and one of the other gods, Kanaloa, were constant companions, Kāne's name was always said first.

Lono was considered god of the clouds, the winds, the sea, agriculture, and fertility. There were over fifty Lono gods worshipped. As god of agriculture and fertility he assumed in one of his forms Lono-i-ka-makahiki, the god of the annual fall festival *(makahiki).*

Kanaloa was best known for his association with his mentor, Kāne; together, they were able to locate water in springs. By some Hawaiians he was considered a god of the seas; he could assume the form of a squid *(mūhe'e).*

The realms and duties of these great gods were not limited to those listed above, and they even overlapped; their "functions" were often shared by lesser gods. The numbers of gods that presided over the activities of mortal Hawaiians, and over one place or another, were countless. Besides the gods that the Polynesians brought with them, other gods were created locally by the Hawaiians themselves. Ancestors were deified; chiefs were deified immediately after death. Spirits *('uhane)* were recalled from the other world by means of prayers and offerings and became special kinds of gods, *'unihipili.* The personal or family gods were called *'aumākua;* some favorites of this kind were owls *(pueo)* and sharks *(manō).*

Those were the invisible gods. Other gods were represented by material objects: idols made of stone, wood, or feathers. Only those made of wood are discussed here.

In their simplest form some of the gods were represented by uncarved pieces of wood; perhaps these were the original, early forms of what later became elaborately carved idols. Skilled craftsmen, devoted to this one craft, made the images according to a technique they had been taught as apprentices, but it is probable that each of them added personal touches, by elaborating on details. The carved figure did not become a god until so converted by the owner or a priest. Thus one god might have different forms.

The carved wooden images *(ki'i kālai 'ia)* had great variation in form and detail, but all had certain general characteristics different from those of images of other Polynesian areas.

The chest was usually prominent, separated from the abdomen by two V-shaped edges, points downward, and adjacent limbs meeting in the middle line. In males the V-shaped edges were a conventional form of the anterior margin of the ribs, and in females they were enlarged to form breasts. The abdomen was usually slightly curved or flat, thus differing considerably from the protruding abdomen of images of central Polynesia. The arms were made free of the body and were pendent, with cupped palms turned inward. In some images hands were pronated, with the forearm and hand rotated so that the palm was directed backward and the thumb next to the body; the thumbs were joined to the outer side of the thighs and the fingers on the outer side. The armpits on the body side were usually defined by anterior and posterior edges that were carried downward on the chest and abdomen. The buttocks were well rounded, with a median cleft, and sometimes exaggerated with a backward projection. In many images genital organs were not represented. The legs were flexed, the thighs exaggerated, and calves greatly enlarged. The feet were usually solid blocks; in some images the front edge was grooved to suggest toes.

Individual features of the head varied greatly in detail: some were decorated with human hair inserted into holes drilled in the normal hair area. In others there was a carved median crest like the crest on feather helmets. Some had an elaborate superstructure erected on the top of the head. Still others had eyes inlaid with mother-of-pearl. The nose was usually a raised ridge expanded at the base to indicate nostrils; in temple images nostrils were elaborately carved. The mouth was elliptical or shaped like a figure eight. In a few images the tongue was made to protrude slightly but never in the exaggerated manner characteristic of Maori carvings. A curious feature in some images was a long lower jaw

Ki'i kālai 'ia

projecting downward and then curving forward with a poorly formed mouth at the end of the chin.

Images can be divided into three general groups: those without supporting props, those with pointed props, and the large temple images, which originally had props that were then cut off below the feet.

The images without props varied in height as well as in certain features: massive upper arms or calves; larger necks than usual; projecting ears; attenuated ears or height; or there were additional features such as in the god Kālai-pāhoa, the sorcery or poison god, which had an elliptical cavity in its back, extending from neck to buttocks, to hold poison.

The images with pointed props *(akua kā'ai)* were carved so that the two feet were together and the remaining wood was trimmed to form a pointed stick, a prop. The prop or support permitted the image to be stuck in the ground or in the house-wall thatch to maintain it in an upright position. In the case of small images the props were longer than the figures, sometimes twice as long. These smaller images were portable. Images with props had the same general body features as those without props with two exceptions: instead of the free pendent position of the hands, these were attached to the thighs; and when a crest was present on the head, it was much more elaborate than in the first type of image.

Akua kā'ai

The large temple gods were essentially the same as the images with props but were much larger. A characteristic of these idols was the flexed knees; sometimes the arms were also flexed. The significance of the flexing of these limbs is unknown. The arms were either free-hanging or the hands were attached to the thighs. As with the images described in the first two types, there were variations from the more typical ones just described. Note is made of the ferocious expression on the faces of many or even most of the temple gods; was this deliberate, to give priests who already had so much power even more, through fear?

The feathered gods *(akua hulu)* are peculiar to Hawai'i, and were probably an outgrowth of the feather helmets. They consisted of a head with a crest of hair, and were truncated at the base of the neck. The foundation of the image was made in the same manner as the helmets, twined with split *'ie'ie* aerial roots. Twining commenced at the top of the head. Hollows, empty spaces for eyes and a mouth, were formed by bending the warps and adjusting the weft rows to form the desired shape. The projecting nose was either woven in place or woven separately and then attached. In the former case a nose was made by forming a triangle with the warps and arranging short bias wefts to widen the lower end to form nostril "flanges." The horizontal weft rows

Akua hulu

were carried from one cheek over the warp projection to the other cheek. In the second, simpler method the nose was woven separately and tied down on the twined background, probably with *olonā* cordage.

The lower jaw angles and chin were made by bending back the warps; twining was then continued to form the neck. The circular neck was extended downward for several inches, with additional bias wefts inserted to increase its diameter at the base. This base was then closed with twining. Usually but not always hoops made from *'ie'ie* aerial roots were inserted inside the image to keep it from sagging.

If the image had a crest, this was made by pushing two warps from the inside through the twined top of the head to form the sides and ends of the crest. With a single-pair twine around the protruding warps, the crest was completed in the same manner as the crests on helmets. In some images there were double crests.

In images with hair instead of crests, tufts of human hair were tied to the twined frame over an area analogous to the hair surface on a human head. Variations included longer hair, braided in the middle with long strands falling to either side of the face; and a median crest of bleached hair placed over the rest of the short tufts of hair.

Feathers were fastened on netting made from *olonā* cordage in the same manner as described for feather capes, cloaks, and helmets. The netting with feathers was then fastened to the twined framework by shaping and tucking it in and around the hollows and protuberances of eye sockets, nose, jaws, and chin, and stitching the feathered net to the framework with *olonā* thread in the same way as was done in making helmets. In the crested images the netting with feathers was closely fitted over the crest; in those with hair, it was fitted along the hair line; in some idols, the netting covered the eye sockets; in others, openings were cut to admit the device that held eyes in place. The netting was usually cut to shape the mouth.

The general color for feathered gods was red, but crests, as in the case of helmets, had yellow feathers on the upper surface and a yellow band around the lower border. In some images, a narrow black band was carried along the side of the crest, continuing down the back.

Eyes, similar to those in some of the wooden images, were made of pearl shell and generally had a wide elliptical form usually set horizontally. However, some of the eyes were nearly round while others were narrow and sometimes set obliquely; some were even crescentic. All had single central drilled holes in the shell, primarily for fixation. The fixation component consisted of a round-headed wooden peg with a long stalk or stem that was fixed to the twined foundation by tying it with *olonā* thread. One method was to tie a piece of the thread, which had been passed through a transverse groove at the inner end of the stalk, to a hoop in the twined framework above the eye. In another method, a piece of *olonā* cordage was threaded through a transverse hole made in the inner end of the stalk, and the two ends tied to a vertically placed stick behind the orbits.

Mouths were either open or closed. When open they were of a simple elliptical shape, or widened at each end, at the corners, to make a shape like a dumbbell or flat recumbent figure eight. A variation of these two more common shapes was a convex upper line and a straight bottom. The technique of making open mouths was the same as for making eye sockets. If the mouth was closed, the lips were twined separately and then attached.

All feathered gods had teeth; usually dog teeth were used, but shark teeth were also utilized. For an image with an open mouth, the crown point of each tooth was filed off and two lateral holes drilled in the remaining tooth. The teeth were then strung on two parallel *olonā* threads through the two holes to form a kind of necklace. This was laid on the lower edge of the mouth with the curves of the teeth pointing upward. A length of split *'ie'ie* aerial root was laid over the surface of the root ends of the teeth in the space between the two threads; the length of root was, in turn, tied down at frequent intervals, between each two teeth, by *olonā* thread passing around the marginal weft row of the twined foundation beneath. The upper set of teeth was fixed in the same manner, with the threaded teeth placed against the upper edge of the mouth and the curve of the teeth turned downward.

When the mouth was not open, teeth were evened by cutting off the roots. Each tooth was tied individually to the inner surface of the upper and lower lips by means of five or six turns of *olonā* thread that passed over the middle of the tooth and under the weft element in the lip flange. Teeth were set closely together.

The nasal aperture and ears were depicted by tufts of black hair or feathers.

SHRINES

A shrine was a simple altar maintained by individuals or small family groups who conducted short rituals not requiring a priest. The ritual was accompanied by an offering, usually a fish or, if that was not available, a taro plant with the same name as the fish, or even a lei. This was laid either on or at the foot of the altar to call upon a god's aid in some particular undertaking. Sometimes the image of the god being called upon was set on the shrine during the ritual, but evidently this was not necessary. All shrines were called *kuahu* except fishing shrines, which were called *ko'a*.

Shrines were of three types: fishing, family, and road. Fishing shrines were of stones and the offerings of nonplant materials, so they are not covered in this book. Family shrines were further divided into two groups: household and occupational. A portion of the men's eating house was set aside or reserved for a household shrine with the family gods placed on it or stuck in the wall beside it. This seems to indicate that only men worshiped at such a "family" shrine, because this house was *kapu* to women. Sometimes a small grass hut was built by the head of a family to keep his special family gods; such a house would be accessible to women. Occupational shrines were located in convenient places for those who tilled the soil, for those who worked in the forests, for fowlers, and for all others engaged in work. These were simple shrines where these men brought their offerings and recited their rituals. Women would have shrines for their special gods.

Road shrines were usually erected alongside paths and trails to well-known areas, a valley for example. Hawaiians believed that various spirits or even definite gods presided over different districts and that the traveler must make an offering to those guardian spirits to ward off any mishap that might otherwise befall him. The shrine consisted of an upright stone or a small stone platform, and the offering might be some leaves, a stone, or, in fact, any material object.

A different kind of roadside shrine was built on the path adjacent to land known as an *ahupua'a.* During the harvest festival known as the *makahiki,* landholders "paid" their taxes by placing offerings, including food, on such shrines. During religious purification ceremonies food offerings such as hard *poi* were placed on these shrines; the officiating priest, after performing the purification rite, ate the food.

TEMPLES

Temples represented expansions of the shrine to accommodate larger groups of people; also, more elaborate ceremonies and rituals were performed in temples. A simple temple consisted of a low, narrow, rectangular platform, facing a paved rectangular terrace. An odd number of upright stone slabs were set at equal intervals along the full length of the rear boundary of the platform.

HEIAU

Complex temples, called *heiau,* varied greatly in ground plans, even on the same island. The great variation resulted from new *heiau* being built frequently; each new professional "architect" *(kahuna kuhikuhi pu'uone)* engaged to plan one would, because of professional pride, attempt to make his *heiau* different from and more outstanding than the others.

Heiau were given individual proper names and were further classified according to the particular function for which they were built: for war, *heiau waikaua,* dedicated to Kū, the god of war; or for the increase of food crops, *māpele,* dedicated to Lono, the god of agriculture. Different kinds of wood and thatching were used in building these two *heiau;* also, different offerings were made in each. *Heiau* were usually built on some dominating site, as on the top of a hill or higher land that overlooked valleys, settlements, or the sea.

An important part of the *heiau* was the court or terrace *(kahua).* The shape, usually rectangular, was sometimes irregular, and a few were even circular. Size depended on the numbers of people to be accommodated. *Heiau* built by kings and high chiefs had large *kahua* to provide room for the many people attending public ceremonies.

Heiau can be divided into those that were terraced, those that were walled, or a composite of both of these. Terraced *heiau* consisted of an open court paved with earth, sand, or large flat stones carefully laid with smaller stones between; there were no boundary walls. Because most terraced *heiau* were built on sloping ground, a stone facing was built on the lower side of the slope so that the court was level. To accomplish this, the area behind the facing was filled with soil, which was sometimes obtained by excavating the upper side of the slope. If the court needed to be enlarged, it was easiest to add another terrace below the first. Still more terraces could be made on down the slope. Steps at one corner led from one terrace to the next.

Walled *heiau* consisted of courts as in the terraced ones but were bounded on all sides by stone walls. Walls were of different types. The most common was made of regular stones piled up with facing stones on either side filled in with smaller stones. Another type was made of parallel rows of large stone facings with the space between filled with rubble. In a third type the outer side of the wall was stepped. Some *heiau* courts were surrounded by walls of closely placed wooden upright stakes.

Set at intervals on the stone walls or part of the wooden fences surrounding *heiau* were carved slabs and/or posts carved with human heads with very high headdresses at their upper ends. These carvings were well shaped; one curious feature was that the chin was carved to extend to form the post below. Often there were rectangular holes with intermediate bars at the head, behind the eyes, to which streamers of tapa were tied.

The composite *heiau* was made with several terraces, some walled and others not; in some of these *heiau* the walled terraces were of greater significance than those without.

Present in many *heiau* was a raised platform located at one end of the court, running across one end and along one side, or at the back center. Such a platform was made by building up sides and filling the space between with earth, rubble, or stones. Other raised platforms built in other parts of the court served as foundations for some of the *heiau* furnishings while others were associated with food offerings.

The main furnishings of a *heiau* were the oracle tower, the houses, the idols, the offering stands, and a refuse pit.

Oracle Tower

The oracle tower *(ʻanuʻu* or *lananuʻu mamao),* unique to Hawaiʻi, was a tall truncated pyramid- or obelisk-shaped structure. Strong timbers were used for corner posts, with the space between covered with rafters of lesser diameter and constructed similar to purlins of houses; however no thatching was applied. Instead, the outside was decorated with white tapa. The oracle tower's name, *lanamuʻu mamao,* came from the names of the three "floors" or platforms that composed the integral parts of the tower: the first or lowest, *lana,* where offerings were placed; the second or central, *nuʻu,* which was sacred and on which the high priest and his attendants conducted certain rituals; and the third or highest, *mamao,* more sacred than the second platform, where only the king and high priest were allowed. It was here that the god of the *heiau* spoke through the priest of coming events.

Houses

Within the court were four houses: *mana, hale pahu, wai ea,* and *hale umu. Mana* was a large house at the farther end of the court. It was the most sacred of all the *heiau* houses because it sheltered the *mōʻī,* lord of all images. There was a great deal of ceremony and many rituals observed in the building of this house. These ended with the tying of a piece of white tapa, *makuʻu,* to the ridgepole; two men dancing a kind of *hula* on the roof; and a priest cutting the thatch over the door, which represented the navel cord. This ceremony was followed by a feast to announce that the house was finished. *Hale pahu* was the house where drums used in *heiau* ceremonies were stored. *Wai ea* was a small house at the entrance to the *heiau* where the king and his counselor, the high priest *(kahuna nui)* conferred. *Hale umu,* built on the other side of the *mana,* was the house in which the fires for the *heiau* were lighted. Outside the *heiau* was a house, *hale o Papa,* where chiefesses worshiped.

The timber most often used to build all these houses was *ʻōhiʻa lehua* or *lama;* in the first case, *pili* was used for thatching, and in the second, ti leaves.

Heiau Gods

There were many *heiau* images; these were described early in this chapter. They were made of *ʻōhiʻa lehua,* with many rites associated with selecting, cutting, and carving them. The task of obtaining the wood was assigned to chiefs of the king's household. As already noted, there were special skilled craftsmen *(kāhuna kālai)* who carved the images. These gods were placed in front of the oracle tower after the latter had been erected. There does not seem to have been any definite number or special arrangement of them.

Stone pavements called *kīpapa* were built on the floor of the court in front of the idols; on these, offerings *(mōhai)* were temporarily placed until a priest was ready to perform the ceremony of presenting them to the gods. Offerings were also sometimes temporarily laid on the ground floor of the oracle tower.

Offering Stands

Wooden stands *(lele)* were erected near the *kīpapa,* and offerings were laid or fixed on the *lele* after they had been presented to the gods by the priests in a ceremony called *hai.* The offerings remained on the stand(s) until they rotted and/or were thrown into the refuse pit to make room for fresh offerings for subsequent ceremonies.

The simplest offering stand consisted of a single pole, probably made of *ʻōhiʻa lehua* because this was the wood used almost exclusively in temples and *heiau,* to which an offering, such as a bunch of bananas, was tied well above the ground. A two-legged support consisted of two slender poles inclined toward each other, often with two horizontally placed boards, like shelves, the lower one of which had holes through which the poles passed; it was on this that offerings were placed. The upper board was fastened at the upper ends of the poles to brace them.

A more elaborate stand had four legs that supported a platform on the upper ends of the poles or a little distance below these ends. In both types the platform was lashed to the poles.

Most offering stands had gourd guards against rats on the poles below the offering platform except in the case of the two-legged type, in which the overhang of the offering board prevented rats from getting to the food.

There were also stone stands that served the same purpose as the wooden ones.

Offerings consisted of pigs, fish, birds, and bananas, usually wrapped in tapa. Flower leis were also used as offerings, with a preference for red flowers, especially *lehua.* Human offerings were also presented.

Luapaʻū

This was a refuse pit into which decayed offering materials were thrown when the stands were needed for fresh offerings. These were usually lined with stones and most were surrounded at ground level with low stone walls.

Types of *Heiau*

The preceding descriptions were of a "typical" *heiau.* Actually there were many different types, each differing in some way from the others. There were the "comfortable *heiau,*" so-called because they did not have the strict *kapu* of some of the others. Others were the *heiau hoʻoulu ʻai,* to increase food crops; *heiau hoʻoulu ua,* to bring rain; *heiau waikaua,* the war heiau; and many more. Finally there was the *heiau* where human sacrifices were offered, *heiau poʻo kanaka.* Here were held purification ceremonies, prayers were said and chanted, the priests fasted, pre-sacrificial offerings were made, and rituals were performed, and, finally, the human sacrifice was prepared.

SOURCES

Buck (1957, 465–532); Ellis (1969, 89–94, 96–99, 164–170); Fornander (1918–1919, 52–159); Kamakau (1976, 129–147); Kirch (1985, 257–260); Malo (1951, 81–85, 159–187); Mitchell (1972, 67–76); Valeri (1985).

12
DEATH AND BURIAL

*M*OE KAU Ā HO'OILO (death) came to the ancient Hawaiians as it comes to all of us today—after an illness, short or long; as the result of an accident; in battle or through some other violent means such as murder. But for ancient Hawaiians there were two other modes of death: becoming a human sacrifice, or being prayed to death by a sorcerer *(kahuna 'anā'anā).*

MOURNING

Upon the death of any individual, except the victim of human sacrifice, a *kapu* was placed upon his or her corpse, on the house occupied by the deceased, on all the relatives who lived in that house, and on their food.

When a king died, the first act was to send the successor of the king, usually an heir, to another district to keep him from becoming defiled *(haumia).* Any person who came into contact with the *kapu* relatives or touched their food also became defiled. Both relatives and neighbors of the deceased gathered before the latter's house and wailed and wept without restraint. Chants eulogizing the deceased were recited. Relatives and near friends showed their grief *(kaumaha)* by cutting their hair in various peculiar ways. Not to cut or shave off one's hair was considered a lack of respect for the deceased and surviving relatives and friends.

Another custom was the knocking out of one or more teeth. This was sometimes done by the mourner himself or herself but more frequently another person did this by placing a piece of hard wood or a piece of a stick against a tooth and hitting the other end with a stone. If a man failed to carry out this "operation," a woman did it while the man slept.

There were other practices to show grief such as tattooing a black spot or line on the tongue with a bamboo sliver dipped in the soot obtained from the smoke of burning cracked *kukui* nuts. All these extreme manifestations of intense grief were known as *mānewanewa.*

Death could call forth a solemn observation in which inhabitants of a whole district came to the house of an *ali'i* in a single file or two abreast, in profound silence, the old in front, with the children bringing up the rear. All were dressed in old fishing nets or dirty pieces of matting to show their grief. As they approached the house they began to lament and wail. One or two were chosen to come forward and chant, reciting, with gestures, the birth, rank, honors, and virtues of the deceased.

The manifestation of grief upon the death of lesser chiefs was less intense, and even less for a commoner. In the latter case the period of *kapu* lasted for only a day or two.

DEIFICATION

Before the body of a king was prepared for deification, it was necessary to determine whether his death was due to sorcery (whether he had been prayed to death by a *kahuna 'anā'anā*). To avoid such a death, *ali'i* carried a pouch upon their person to hold their personal hair trimmings and nail parings, to keep these "baits" *(kāmehaʻi* or *maunu)* from being acquired by a sorcerer who could use one of these materials in praying the victim to death.

The body of the king was removed to the men's house *(mua)* and placed upon or beside the shrine, where the ceremony accompanying deification was conducted by a special priest, *kahuna hui.* Except in rare cases there were no deification ceremonies for chiefs and commoners.

Once the deification ceremonies were over and the body properly disposed of, the *kapu* was lifted and the king became worshiped as a god. His successor returned from his "exile" and built a special *heiau* to contain the deceased king's bones if that was to be their disposition.

PREPARATION OF THE BODY

There were several ways in which a body was made ready for burial. In one such method the body was first wrapped in leaves of banana, *wauke,* or/and taro; this rite was known as *kapa lau,* garment of leaves. The covered body was then placed in a shallow pit and a fire lit over it. The fire was kept burning for ten days, during which a purification prayer *(pule huikala)* was constantly recited. When the flesh had separated from the bones, the latter were gathered together and wrapped in tapa; the bundle was deposited in a temple or elsewhere as noted. The flesh and intestines *(pela)* were thrown into the sea at night, burned, or buried.

When the body as a whole was to be preserved, another method of preparing the body for burial was used. A transverse cut was made just below the rib cage, the viscera removed, and the cavity filled with salt to preserve the body. This was especially important when the body of an *aliʻi* had to be preserved for ten days. Sometimes, after the viscera were removed, the body cavity was filled with the fawn-colored silky hairs *(pulu)* that cover the young furled fronds of the native tree fern, *hāpuʻu pulu,* to absorb the body fluids. A body so treated was called *iʻaloa* (long fish). The body was then wrapped in tapa—an inner thin sheet with one or more thicker sheets in outer layers; sometimes the tapa bundles were further wrapped in *lau hala* or *makaloa* mats.

Although apparently not mandatory, the bodies of kings, chiefs, and sometimes priests were buried in the extended position, those of commoners in the flexed position. In the latter case, before rigor mortis set in, the head was bent forward toward the knees, the knees were drawn up toward the chest, the hands were placed under the buttocks and then the head, hands, and knees were bound together with cord.

CARRYING THE BODY

There were various ways in which a body was carried to the place of burial. The simplest and probably the manner in which the bodies of commoners were carried was to sling the body to the middle of a carrying pole with strong bands of tapa or *hau* cordage. The body of a king, and probably that of a chief, was carried in a more dignified manner by means of a kind of litter or stretcher *(mānele)* made of two carrying poles with short crosspieces lashed to them. The body was placed on the litter, which was carried either by two men, an end of each pole on a shoulder, or by four men, each with a pole end resting on a shoulder.

PLACES OF BURIAL

The place of internment for a king, if outside a temple, was usually a well-concealed secret cave known only to a trusted retainer *(kahu)* whose position with the family was usually hereditary, and to selected friends who were expected not to divulge the secret site of burial. When those who had carried the body returned from the burial site, they washed themselves in fresh water and seated themselves in a row. A priest sprinkled them with salt water from a calabash that also contained seaweed—this is a ritual associated with purification on various occasions. Then a ritual chant was recited to free them from the *kapu* placed upon them for having carried a god, the deified king. It seems more likely, with the possibility of the secret being revealed through torture, that those involved in carrying the body to the burial site were put to death—this would not have been too great a sacrifice to make for their king.

A royal mausoleum, *ilina o nā ali'i,* might also serve as a temple as well. One of these was the Hale-o-Keawe near the City of Refuge, Hōnaunau. This was built in 1740 by Kanuha for the bones of his father, King Keawe, great grandfather of Kamehameha I. This mausoleum/temple consisted of a thatched house placed in a square paved with stones and surrounded with a fence made of thick wooden stakes and planks. Outside the fence were arranged carved wooden images. At one side of the square, inside the fence, was a stage of tree trunks. Inside the house were carved wooden and feathered idols, an altar on which all kinds of offerings were laid, and the bones of the king wrapped in tapa.

Chiefs were sometimes buried in the *mana,* the temple house in a *heiau,* or near the middle of the temple court of the *heiau.* Sometimes a priest might also be buried in the latter site.

Many bodies of commoners were buried beneath dwelling houses.

At times, a man or woman secretly exhumed the body of a loved spouse, removed the four long leg bones and skull, and washed them in water to clean them. The bones were then wrapped in tapa and taken to bed and slept with at night. The number of bodies treated in this manner was considerable among those who were fond of one another.

Probably sand burial was the most common. Besides mass burials of warriors on battlefields there were multiburials of men, women, and children in sand on all the Islands.

Some bodies were buried in earth, often near the dwelling house of the deceased. Such graves were usually marked with a pavement of stones or outlined with a border of stones.

Burial in caves, already mentioned, was also common. All the Islands have many cliffside caves, the result of volcanic action. Caves were used for the burial of both the nobility (already noted) and commoners, differing only in location, size, and "furnishings." A chiefly family had its own cave "vault." Some entrances to caves were small so that the location of the caves was easily concealed and protected. When the entrance was large, these were blocked with stone walls. It was not difficult to approach a cave near the base of a cliff, but those higher up the face of a cliff presented problems. Narrow paths were made; these were dangerous even when first constructed. They were either deliberately destroyed or, with time, became obliterated by subsequent weathering and/or landslides. There is also the possibility that a body and its carriers were lowered from the top of the cliff with strong cords.

Sometimes a body was placed in a half canoe that served both as litter, carried on the shoulders of bearers, and bier. The cut end of the canoe was either left open or was closed. The body was flexed and wrapped in tapa before it was placed in the canoe on its back; a moss "pillow" was placed beneath the head. Such a burial was in a cave or lava tube.

Burial in cists was not common. It was once thought that stone cists were exclusively for the burial of priests, but that is not so. The sides and ends of cists were lined with two rows of stones, one above the other; there were no stones on the bottom, the floor. The body was placed on the floor in the extended position with the head facing upward; it was then covered with earth. Some cists were left unmarked, others had two short rows of small river pebbles placed across the ends and five parallel rows of the same pebbles in the long direction. Such burials may be of postcontact origin.

Today, platform tombs are sometimes mistaken for house platforms, which they superficially resemble, and vice versa. Covered with stone platforms, they were excavated earthern chambers, most of which were rectangular. The middle portion of the platform was paved with smaller stones and coral; beneath this portion were two slabs of coral. Below these slabs was an earth chamber in which bones and the skull, wrapped in tapa, were laid. Sometimes the excavation was lined with stones and was vaulted with tree trunks. When the latter were present they supported a low platform paved with coral.

Of special interest are the twined sennit caskets *(kā'ai)* used to hold the bones of certain kings; they are extraordinary and peculiar to Hawai'i. They had a head, with defined features; a neck; and a somewhat cylindrical body without arms and legs. The neck was short and expanded into fairly square shoulders. The trunk was widest at the shoulders, diminished slightly toward the middle, and expanded again at the base where it was truncated and circular. The ears were fan-shaped with an outer convex curve; they lay fairly flat against the head. The nose was a long projecting ridge that widened at the lower, nostril end. The eye openings were made by not crossing the horizontal wefts to interweave with the vertical warps. Elliptical pieces of pearl shell covered the eye openings, held in place with a round-headed wooden pin that also served as a pupil. The mouth was open and made in the same fashion as the eye openings.

The material used to make these caskets was five-ply sennit. The technique used was a form of finger weaving, quite distinct from the twining used in making some receptacles and feather helmet and image frames. A simple horizontal weft was used to make continuous spiral turns over and under alternate warps; these latter radiated vertically from a commencement center. Because the weft was continued in spiral turns, the technique might be called circular weaving or twining. The workmanship on these caskets is truly beautiful.

BELONGINGS OF THE DECEASED

Various belongings of the deceased were placed where he was buried. Such materials were extensive and of great value in the case of nobility: fine tapas and mats, personal ornaments, spears, images, and others. There was less in quantity and value of artifacts buried with lesser chiefs and even less with commoners. These articles were buried with the body of the deceased, or with his bones, because they were associated with him and were therefore *kapu.* As in many other cultures, food offerings were also placed with the body or bones.

SOURCES

Buck (1957, 564–580); Ellis (1969, 139, 164–166, 174–182, 357–367); Fornander (1919, 570–577); Handy and Pukui (1972, 115, 146–159); Kamakau (1964, 33–44); Kirch (1985, 71, 111, 113–114, 161, 167, 169, 204, 237–246, 253); Malo (1951, 96–107); Pukui et al. (1972).

DESCRIPTION OF PLANTS

PLATE 1
'A'ali'i
Dodonaea viscosa Jacq.
(indigenous)

'A'ali'i ranges from shrubs one foot high to trees up to thirty feet high. Leaves are alternate, short-stemmed, narrow, more or less spatula-shaped, blunt or pointed, one to four inches long and 1/4 to 1-1/4 inches wide. Flowers are inconspicuous, borne in small clusters on the tips of branches or in leaf axils. They are normally unisexual, with the two sexes usually borne on separate plants. The fruit is a yellow, red, or brown capsule about 1/2 inch long with two to four broad-winged angles; it contains two to four cells, each with one or two ovate seeds. The wood is hard, yellow-brown, with a dark heartwood.

PLATE 2

A'e or *Mānele* (soapberry)

Sapindus saponaria L.

(indigenous)

Although *a'e* is usually a small tree, it may reach a height of eighty feet at higher altitudes (e.g., at 4,000 feet). It is native to Mexico and South America, and west to Hawai'i and a number of islands and to Africa. The foliage is dense and broad; the leaves are compound, light green, with three to six paired or nearly paired narrow, pointed, and somewhat curved leaflets, two to five inches long and unequal at the base. Usually the axis of mature leaves is slightly winged; thus the name "winged leaf" is sometimes applied to this plant. Large clusters of tiny, unisexual, five-parted flowers appear at the end of branches in their season. Fruits are single and round, or two- to three-lobed; when single, this is accompanied by two undeveloped lobes. The fruit is about 3/4 inch in diameter, brown-skinned with shiny, yellowish, soapy pulp; this surrounds a large, globose, brown or black seed.

PLATE 3
'Ahakea
Bobea spp.
(endemic)

'Ahakea are native Hawaiian trees belonging to the coffee family. They have oppositely arranged, small to medium-sized oblong leaves that are often pale or yellowish green in color. Their small, greenish to white tubular flowers occur in axillary clusters of one to three. The small purplish black juicy fruits, which are ovoid to globose, contain two to twelve elongate seeds. The wood is yellow and very hard.

PLATE 4

'Āheahea or *'Āweoweo* or *'Ahea* or *Alaweo*
Chenopodium oahuense (Meyen) Aellen
(endemic)

'Āheahea is a native Hawaiian shrub belonging to the goosefoot family. It very much resembles a close relative, the weed called lamb's quarters. The leaves are simple, ordinarily alternate, and without stipules. The flowers are small and arranged in panicles. The leaves are thick and narrow to triangular or rhomboidal. The flowers are usually perfect, without petals, and ordinarily symmetrical. The fruit is a nut.

'Ahu'awa
Mariscus javanicus (Houtt.) Merr. & Metcalf
(endemic)

'Ahu'awa is a native Hawaiian sedge. It grows in wetland areas such as marshes. The plant is one to four feet high; at its base is a tuft or cluster of long, rough, narrow, pale green leaves. A radiating inflorescence is borne on a slender stem that is longer than the leaves. The inflorescence is surrounded by five or six rough leafy bracts.

'Aiea
Nothocestrum spp.
(endemic)

'Aiea are native Hawaiian plants that belong to the nightshade family. They are soft-wooded shrubs or small trees with leaves that are ovate or oblong in shape. Their yellowish or greenish yellow tubular flowers are nearly stemless and are borne in compact axillary clusters that emerge from knobs along the stem. The sepals, which are sometimes as long as the floral tube, are usually brownish-hairy. The fruits are fleshy, ovoid, globose, or elongate and yellow to orange (rarely whitish).

'Aka'akai or *Nānaku* or *Naku* (great bulrush)
Schoenoplectus lacustris (L.) Palla
(indigenous)

'Aka'akai is a sedge that grows on the edge of fresh- and brackish-water marshes. The plants have unbranched, slender, dark green spongy stems, three to nine feet high, with basal diameters as great as one inch. The stems are erect, rising from a strong, creeping underground stem. There are a few short, leaflike sheaths at the bases of the stems. When not in flower, this sedge looks like a giant onion plant. When in flower, the stems bear pointed leaflike bracts at their tips; this subtends a "head" consisting of a compound cluster of reddish brown flowers. The latter are stemless or are borne on stems about 2-1/2 inches long. Individual flowers are minute and have six minute, spiny bristles.

PLATE 5
'Ākala
Rubus hawaiensis A. Gray and R. *macraei* A. Gray
(endemic)

The name *'ākala,* meaning pink for the color of the juice of the berry, is given to two species of Hawaiian raspberry, *Rubus hawaiensis* and *R. macraei.* The former has erect canes (stems), five to fifteen feet high, with prickles. Leaves are about six inches long, have three leaflets, or are three-lobed or undivided, with petioles about an inch long. The latter-named *'ākala* is a large trailing shrub with the stems bearing wooly reddish yellow hairs and red prickles; leaves are similar to those of the other species. Both species have flowers with a hairy, deeply five-parted calyx about 1/2 inch long, and dark pink to dusky rose petals. Fruit of the two species are red to dark purple; they are both edible. A rare form of *Rubus hawaiensis* has yellow fruit.

PLATE 6

ʻĀkia or *Kauhi*

Wikstroemia spp.

(endemic)

ʻĀkia are native Hawaiian shrubs or trees with jointed branches that repeatedly fork in pairs. The leaves are opposite, or subopposite if a slight displacement of one of a pair occurs; they are short-stemmed, small, oval or narrow, smooth on both sides, and with smooth margins. The flowers are tiny, yellowish, four-parted, and tubular, without petals. They are borne in clusters at the tips of branches or in the axils of leaves. They may be perfect or unisexual. The fruits are ovoid drupes; they are small, from 2/10 to 1/2 inch in diameter; they are short-stemmed. Their color ranges from yellow to orange to red. A juicy, thin-skinned pulp surrounds a single seed.

PLATE 7
'Akoko
Chamaesyce spp.
(endemic)

'Akoko are native Hawaiian shrubs or trees, some species of which occur along the coast, some in the lowlands and foothills, others in wetland or dryland forests. A few are found in the uplands above 5,000 feet. They have shiny dark green leaves, three to five inches long and one inch wide. They bear a number of ovoid, three-angled, green, red, or pink fruiting capsules at the tips of branches. The capsules are up to an inch long. These plants get their name from the color of the capsules: *koko* is blood and *'akoko* means blood-colored. Plants are characterized by the milky sap.

PLATE 8
ʻĀlaʻa
Pouteria sandwicensis (A. Gray) Baehni & Degener
(endemic)

ʻĀlaʻa is a large native Hawaiian tree with smooth oblong leaves that are shiny green on the upper surface and bronze on the underside. The flowers are nearly stemless or long-stalked and occur singly or in clusters of two to four in the axils of the leaves. The fruit is globose, ovoid or pear-shaped, purplish black (yellow in one form), and up to two inches long. The tree has a sticky, milky sap. The wood is hard.

Plate 9

Alahe'e or *Walahe'e* or *'Ōhe'e*

Psydrax odorata (G. Forster) A. C. Smith & S. Darwin

(indigenous)

Alahe'e grows as a large shrub or small tree, two to twenty feet high, in dry regions, often on lava flows. The wood is very hard. Its oblong leaves, from two to three inches long, are dark green and shiny. Clusters of flowers develop at branch tips; they are four- or five-lobed and about 1/4 inch long; they are white and fragrant. The fruit forms an inverted ovoid, about 1/3 inch long, and is black.

PLATE 10
Aloalo (native hibiscus)
Hibiscus spp.
(endemic)

These plants are usually shrubs having smooth or rough leaves that are scattered along the branches or at the tips only. The flowers are borne singly near or at the tips of branches. The calyx is a five-lobed cup, accompanied by few to several bracts. There are five petals, varying in size and color—from white to yellow to orange to pink to red. In some species there is a dark patch at the base of petals. The long staminal column is more or less covered with stamens and surrounds all but the tip of an even longer five-lobed style. The fruit is a capsule containing fifteen or more seeds.

PLATE 11
'Ama'u
Sadleria cyatheoides Kaulf.
(endemic)

This native fern grows in the uplands, on open lava flows or in wet forests; it has a short trunk, about one to two feet, rarely to nine feet, high. The young fronds are a bright red and turn green with maturity. Like the *hāpu'u* ferns, the fronds are made up of many segments; these are uniformly light green with a single prominent midvein.

Plate 12
'Ape
Alocasia macrorrhiza (L.) Schott
(Polynesian-introduced)

This plant is a close relative of taro and resembles that plant in some respects. *'Ape* has huge heart-shaped, shiny, light green leaves with blades four feet long and 1-1/2 feet wide; these are attached to petioles up to 4-1/2 feet long. The leaf tends to point upward in contrast to taro whose leaves point downward. The tuber/roots of this plant needs to be prepared in a special manner (by boiling and repeated decantations) to make them fit for eating; they are otherwise inedible because of a high calcium oxalate content.

PLATE 13
'Awa
Piper methysticum G. Forster
(Polynesian-introduced)

'Awa is a shrub belonging to the pepper family. It is from four to twelve feet tall, with either all green, green with dark spots, or blackish purple, jointed stems that are swollen at the joints (nodes). The leaves, with inch-long petioles, are arranged alternately on the stem; they are large and heart-shaped, five to eight inches long and nearly as wide, with eleven to thirteen prominent veins radiating from the leaf bases. Flowers are borne in short, narrow spikes; the two sexes are reported as occurring on separate plants. All parts of the plant contain a relaxant or sedative element.

PLATE 14

'Awapuhi or *'Awapuhi kuahiwi* (shampoo ginger)
Zingiber zerumbet (L.) Sm.
(Polynesian-introduced)

'Awapuhi is a common plant in the lower parts of damp open forest. It has a lush growth of a dozen or more thin leaves up to a height of a foot or two. The leaves are four to eight inches long and two to three inches wide; they are smooth on the upper surface and more or less hairy on the underside. These arise from large, knobbed, aromatic underground stems (rhizomes) which grow horizontally and lie near the surface of the ground. In late summer, an oblong inflorescence, two to five inches long, appears on a stalk about a foot high, separate from the leaf stalks. The flower head consists of large, greenish red overlapping bracts covering small inconspicuous yellowish flowers, of which only one is open at a time. The mature flower head is saturated with a sudsy, slimy, highly aromatic sap.

Plate 15
'Āwikiwiki or *Puakauhi*
Canavalia spp.
(endemic)

The name *'āwikiwiki* is applied to various species of *Canavalia,* members of the pea family, most species endemic to a different island in Hawai'i. They are vines with densely tawny young shoots; the leaves consist of three broad leaflets with pointed tips. The pods are comparatively narrow, about five inches long and one inch wide, and smooth or comparatively so. The seeds are flat and brown; some may have black mottling. Flowers are magenta to purple.

Plate 16

Hala or *Pū hala* (screw pine)
Pandanus tectorius S. Parkinson ex Z
(indigenous?)

Hala is a tree with wide-angled branches that grows twenty or more feet high, usually at lower altitudes in dry areas. It either is indigenous to Hawai'i or was brought here by the Polynesian settlers. It is characterized by its few to many prop roots: straight, cylindrical, fibrous aerial roots *(ulehala)* that issue from both the trunk and the branches. These roots eventually enter the soil and function as regular soil roots. While aerial roots, their tips are protected with a root cap consisting of layers of paperlike membranes. Parallel rows of short spinelike projections are present along the length of these roots; they are rudimentary secondary roots or rootlets. Sometimes the trunk is gone and the tree crown is supported entirely by the prop roots. The prop roots are essential to the tree because they support the large, heavy clusters of leaves and fruits borne on the end of the branches. The leaves, called *lau hala,* are spirally arranged at the branch ends. They are three feet and more—up to six feet—long, and two to six inches wide. Their bases are wide and flaring, clasping the circumference of the branches, with their tips sharply pointed. The scars left when leaves drop appear as closely placed rings on the trunk and branches. The leaves bend at about midpoint and drop at right angles. Small hooked spines *(kōkala)* occur on the margins and on the dorsal (under) side of the midrib *(kuakua).* Male and female inflorescences are borne on separate trees. The female inflorescence consists of a roundish cluster of many appressed flowers that develop into fruitlets to make a globose or ovoid composite fruit *('āhui hala)* about eight or more inches long. This fruit consists of fifty or more angular, wedge- or cone-shaped yellow to red drupes, popularly known as "keys" *(hala),* about two inches long and one inch wide, attached to a "core" *('īkoi hala).* The inner, fleshy end of the drupe *(pua hala)* is fibrous while the outer end *(iwi hala)* is woody and contains four to twelve one-seeded or empty cells. The male inflorescence *(hīnano)* consists of a "spike" about a foot long surrounded along its length by overlapping white, narrow, pointed, fragrant bracts that subtend (cover) large clusters of very small flowers conspicuous for their pollen *('ehu hīnano).*

PLATE 17
Hala pepe
Pleomele spp.
(endemic)

Hala pepe, a type of Hawaiian dracaena, are native plants belonging to the agave family. They are woody, treelike plants, reaching heights of twenty or more feet in dryland or mesic forests. They are much-branched about a straight trunk. The long, narrow leaves lack petioles and are clustered at the ends of branches. The flowers are about two inches long and hang in clusters from the branch ends as well. The fruits are round and yellow or golden.

PLATE 18
Hāpu'u (tree ferns)
Cibotium glaucum (Sm.) Hook. & Arnott and *Cibotium menziesii* Hook.
(endemic)

Hāpu'u are native ferns—the most common tree ferns in Hawai'i. Their abundant growth forms forests in both dry and damp regions. In other areas, they are associated with *'ōhi'a lehua* trees. There are two species called *hāpu'u.*

Hāpu'u pulu (Cibotium glaucum) develops a large trunk, from six to eight to over twenty feet high. The trunk, one to three feet in diameter, consists of dark-colored interwoven roots that surround a small central stem, three to four inches in diameter. The center (core) of the stem, filled with starch, is enclosed in a hard, almost glass like, black layer. An older stem can contain as much as sixty to seventy pounds of starch. At the top of the trunk is a cluster of fronds, each six to twelve feet long. The frond stems are brown and smooth. The broad fronds are intricately divided into hundreds of small segments, termed pinnae. These are dark green and shiny on top, and a lighter green or dull white and may be coated with fine cobwebby hairs on the underside. On the edges (margins) of the undersides are spore cases filled with spores.

The description given for *hāpu'u pulu* covers *hāpu'u 'i'i (Cibotium menziesii)* as well. The two species have different types of *pulu,* a kind of "wool" that covers furled youngest fronds (frond "buds") and the bases of young unfurling and older fronds at the top of trunks. These hairs *hāpu'u pulu* are soft, silky, and yellowish brown, but in *hāpu'u 'i'i* they consist of coarse, stiff, reddish brown or black bristles.

PLATE 19
Hau
Hibiscus tiliaceus L.
(Polynesian-introduced?)

Hau is a member of the mallow family and is related to the ornamental hibiscus. It was brought by the Polynesians that settled Hawaii, although some botanists believe it also to be indigenous. It is a much-branched tree, twelve or more feet high. Although commonly a tree of the lowlands, it can also be found at higher elevations. There are two varieties: a rare erect one, *hau oheohe,* which was chiefly grown for its bast fibers, and a creeping variety that grows gnarled and crooked, of low to medium height, often spreading horizontally over the ground. It grows in thickets, forming an almost inpenetrable network of trunks and branches. The leaves are large, rounded heart-shaped, varying in size from two to twelve inches in diameter; they are leathery to touch, and their margins are entire or shallowly scalloped. On the upper surface they are smooth and shiny but on the underside they appear dull (whitish because of the presence of matted hairs). Flowers grow profusely at and near the tips of branches; they have the appearance of the flowers of present-day cultivated single hibiscus. The petals appear as light to bright yellow "cups" from two to three inches deep and four to five inches across the "mouth" of the flower. Most forms have dark red centers. As the day progresses, the color of the petals changes to a dull red or mahogany. By the next morning the petal complex, (the corolla) has fallen to the ground, still fully "open," leaving the calyx and ovary on the branch. The calyx cup is an inch long, five-toothed, downy, and persistent; ten to twelve shorter bracts surround (are exterior to) the calyx. The fruit is downy, ovoid capsule, about an inch long, with each of the five valves containing three smooth seeds.

Plate 20

Hinahina or *Hinahina kū kahakai*

Heliotropium anomalum var. *argenteum* A. Gray

(indigenous, the var. endemic)

Hinahina, a member of the borage family, is a native of many islands in the Pacific as well as of Hawai'i. It grows on dry sandy beaches not far above the high-tide mark. The plants are perennial herbs, growing prostrate with the tips ascending a few inches. Both the branches and leaves are silvery as a result of being covered with silky, flat-lying hairs. The leaves are narrow and about an inch long; some widen near the tip. The flowers are white or rarely purplish with a yellow center and are fragrant. They are borne in one-sided coiled spikes on forked stalks. The calyx is tiny with the funnel-shaped corolla being slightly longer; the latter is about 1/4 inch long.

PLATE 21
Hōʻawa
Pittosporum spp.
(endemic)

Hōʻawa are native Hawaiian trees with leaves that are shiny and wrinkled on the upper surface and smooth or covered with dark to light brown woolly hairs on the underside. They are from four to ten inches long and are broadest above the middle, where they are one to two and a half inches wide. The flowers grow in short clusters in the leaf axils and along the stem or, rarely, at the branch tips; they are about 1/3 inch long, cream-colored and unisexual. The fruits are large—generally two to three inches long although some species have fruit only an inch long and somewhat less in width—and thick. The outside appears woolly when the fruit is immature but becomes smooth when the fruit matures. The fruit, a capsule, is two- to four-valved; the inside surface of the capsules is orange; the pulp is generally viscid. Seeds are black and arranged in two rows.

Plate 22
Hoi (bitter yam)
Dioscorea bulbifera L.
(Polynesian-introduced)

The leaves of the so-called poisonous yam, *hoi,* are lighter on the underside and are heart-shaped with an elongated, very pointed tip; they are arranged alternately on the stem, which is round in cross section. Small roundish aerial tubers form in the axils of the leaf stems; these have an ashen gray skin, with a thin, dark green rind beneath the skin. The flesh is yellowish green or greenish yellow. The underground tuber is usually roundish, but may be a flattened "sphere," or even fan-shaped. The tubers are reddish brown on the outside and entirely white on the inside.

PLATE 23
Hōlei
Ochrosia spp.
(endemic)

Hōlei, members of the dogbane family, are native to Hawai'i. They have oval or inverted ovate, long-pointed, leathery leaves, up to six inches long; these are arranged in pairs, threes, or fours. The leaf blades taper at the base into leaf stems about 1/2 inch long. The flowers are small, cream-colored, fragrant, and stemless. They develop in short clusters at or near the branch tips. Fruits are yellow to purple, twinned, ovoid or ellipsoid, pointed drupes, each rather small. Sap in the bark and roots is white.

PLATE 24
'Ie'ie
Freycinetia arborea Gaud.
(indigenous)

'Ie'ie is a native Hawaiian vine that belongs to the same family as *hala*. Commonly it grows in the forests from altitudes of 1,000 to 4,500 feet. It climbs to the top of trees and forms a luxuriant, impenetrable cover on the ground below. The stems are ringed (scars of fallen leaves) and are about an inch in diameter. Branches form every few feet; these end in tufts of narrow spiny leaves that may be up to 2-1/2 feet long and up to two inches wide at their base, tapering to a point. An inflorescence consisting of two to four cylindrical spikes arises from the center of the leaf cluster. The spikes are about five inches long, with a diameter of 1/2 inch. They are surrounded by leafy, edible bracts that are rose-colored on the undersurface. The fruit is a spike crowded with orange, pulpy, many-seeded berries. Many long, narrow, uniform-sized, strong, and pliable aerial roots emerge along the stem. These may be as long as twenty feet, with some eventually striking into the soil and becoming soil roots.

PLATE 25
'Iliahi (sandalwood)
Santalum spp.
(endemic)

'Iliahi are native evergreen shrubs or trees. They partly depend on other plants for food; special structures on their roots extract nourishment from the roots of adjacent plants. Their leaves are opposite, feather-veined, up to four inches long, and dark glossy green or smaller, somewhat thick, and pale green. The flowers, which are perfect, form in panicles at the end of branches or in the axils of leaves. They have a four- or five-lobed red or green or greenish yellow calyx; the fruits are drupelike. The heartwood is scented.

Plate 26
'Ilima
Sida fallax walp.
(indigenous)

'Ilima, a member of the mallow family, is native to Hawai'i. It grows from near sea level to more than 6,000 feet. It takes various forms, from shrubs about ten feet tall to almost postrate. The leaves vary from oblong to heart-shaped with tips pointed or blunt. The edges of the leaves are scalloped; they are an inch or more long. The flowers range from yellowish to orange, sometimes reddish brown at their bases. They are about an inch across and are five-parted. They generally grow solitary or in twos or threes together, or multi-flowered, on stalks near the tips of branches. The calyx is ten-ribbed and downy; it is about half as large as the corolla. The fruit is a wide, seven- to twelve-spoked, wheel-like, one-seeded, beaked seed case or carpel. All parts of the plant except the corolla are usually covered with a white down. Some mountain forms are glabrous.

PLATE 27
Kalo (taro)
Colocasia esculenta (L.) Schott
(Polynesian-introduced)

Taro is a perennial herb consisting of a cluster of smooth heart-shaped leaves *(lau kalo)* rising a foot, and often more, from an underground stem *(kalo),* botanically known as a corm. The flesh of the corm is dense and visually homogeneous ranging in color from white, yellow, lilac-purple, and pink to reddish. Leaf blades may be all green or have variegated stripes or spots; some are white or dark purple mottled. The leaf stems (petioles), called *hā,* are a little longer than the blade and are attached peltate, away from the edge. The point of attachment of the leaf stem to the blade is called a *piko* because of its resemblance, on the upper surface of the leaf, to a human navel. *Hā* may be green, green and white, dark purple, pink, or reddish. On the upper side of the *hā* is a furrow *(māwae);* in this groove or furrow of the last expanded leaf lies furled the youngest (the last formed) leaf, protected there before it emerges and expands. There are prominent air passageways in the petiole. The top of the leaf stem is bent so that the tip of the blade points down. The inflorescence consists of a spadix *(ʻīkoi pua),* a spike covered with flowers—sterile at the tip and middle, fertile male (staminate) flowers between these two groups, and fertile female (pistillate) flowers at the base. The spike is enclosed in a creamy white or yellow bract, bontanically known as a spathe, constricted below the middle. Seeds are produced, but rarely. All parts of the plant have special cells, larger than those in the surrounding tissue; these special cells contain bundles of fine, needlelike crystals, known as raphides, of calcium oxalate; these make for an acrid reaction when any part of the plant is eaten uncooked. When "raw" plant material is taken into the mouth and chewed, the pressure of chewing and the saliva in the mouth loosen and break up the cell walls so that the raphides are ejected into the lining of the mouth and the tongue. The Hawaiians call this irritating sensation *maneʻo* (itchy). It had always been difficult to attribute the extreme pain caused by these needlelike "projectiles," but since the advent of the electron microscope, with its ultramagnification, we now know that these are not simple,

smooth-surfaced needles but are barbed, which explains their great pain-producing effect. The underground stem bears buds (*'ohā*) that develop into suckers, also called *'ohā*. The number of these borne by a single corm varies with the variety and/or growing conditions.

Plate 28
Kamani (Alexandrian laurel)
Calophyllum inophyllum L.
(Polynesian-introduced)

Kamani is a handsome, low-branching tree, up to sixty feet high. The bark is rough and gray; the wood is hard and tough. The leaves are nearly oval, somewhat shiny, leatherlike, and from three to eight inches long. A feature of the leaves is the fine, closely placed parallel side veins that form a unique and pleasing pattern. The flowers are white, an inch across, and are borne in clusters of four to fifteen. The flowers are perfect, with four sepals, four to eight petals, and many bright yellow stamens. The flowers are very fragrant; in appearance and scent, they are reminiscent of orange blossoms. The fruit is globose, about an inch or a little more in diameter. It has a thin leatherly skin that is yellowish green; this skin covers a light-colored bony shell that in turn encloses a single kernel surrounded by cork. The kernel is white and very oily.

PLATE 29

Kauila or *Kauwila*

Alphitonia ponderosa

Colubrina oppositifolia

(endemic)

The name *kauila* or *kauwila* is given to two different native Hawaiian trees, both members of the buckthorn family: *Alphitonia ponderosa* and *Colubrina opositifolia.* The former is found on all six of the main Hawaiian islands. It has alternate, thin, oblong to narrow leaves that are grayish or rusty-woolly on the underside; petioles are up to an inch long. The wood is hard and red-and-black streaked. The latter species has opposite, thin, pale green, ovate or oblong leaves that may be up to seven inches long; petioles are up to two inches long. A conspicuous gland is found at the base of each vein on the underside of the leaves. The wood is hard (even harder than the first-mentioned *kauila*), heavy, and dark red. *Colubrina oppositifolia* is only found on O'ahu and Hawai'i.

PLATE 30
Kauna'oa
Cuscuta sandwichiana Choisy
(endemic)

Kauna'oa is a parasitic vine, a species of dodder, with slender yellowish or orange stems and branches. The vines bear suckerlike roots by which they attach themselves to other plants from which they obtain their nutrition. Flowers are small, the corolla yellowish and five-lobed to the middle. Fruits are capsules containing small, dark reddish seeds.

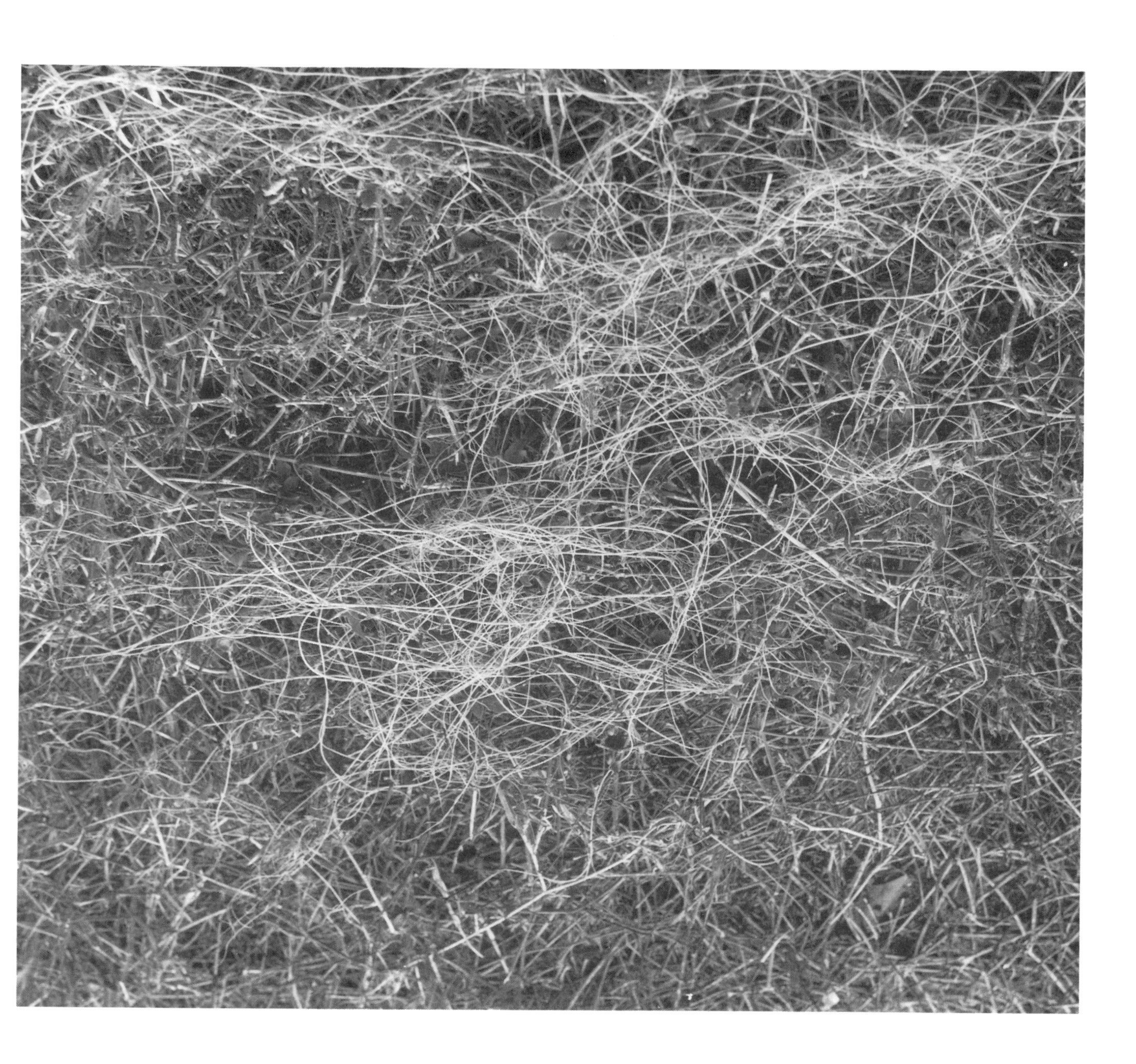

PLATE 31
Kī (ti)
Cordyline fruticosa (L.) A. Chev.
(Polynesian-introduced)

Ti is a simple, sparingly branched plant three to ten or even more feet high. The stem is slender, upright, and marked at short intervals with leaf scars. The leaf is from one to two, and up to almost four feet long, and four to six and even more inches wide. It has a prominent midrib and a long (two to six inches) leaf stem or stalk (petiole) that is deeply grooved. The leaf is light green, with the upper surface shiny. The leaves are closely arranged on the stem in a spiral, concentrated at the apex in a "feather-duster" cluster because lower leaves are continually turning yellow and dropping. The root is long and fairly thick; it may weigh up to 300 pounds. It is fibrous and filled with the reducing sugar fructose. The plant flowers in spring; the inflorescence (flower cluster) is pendant, branched, and about one foot long. It bears numerous closely set flowers that are complete or bisexual. They are about 1/2 inch long, white touched with pale purple. Upon opening, the petal lobes curl back, exposing six yellow stamens and one white pistil. Small berries, changing from green to yellow to red, develop.

PLATE 32
Kō (sugarcane)
Saccharum officinarum L.
(Polynesian-introduced)

Sugarcane is a giant grass; its stems, commonly called "stalks," are coarse, thick, normally unbranched, six to fifteen feet tall or even taller, and up to several inches in diameter, with short conspicuous internodes. The skin or "rind" varies in color with variety: yellow, yellow-and-pink striped, reddish, dark purple, or green with white stripes. The stem is filled with a solid, fibrous, juicy pulp. The sweet juice or "sap" was the chief purpose for growing this plant; the Hawaiians did not crystallize sugar as we do today. Leaves measure from one to several feet in length, and vary from one to two or more inches in width; they are smooth except for their margins, which are serrated (saw-toothed). The leaves on the lower part of the stem die and hang down; the upper leaves are almost vertical and loose-clustered. In Hawai'i, the inflorescence, commonly called a "tassel," forms at the upper end of the stalk about November. The inflorescence varies in length from one to two feet and is multibranched. These branches bear many small flowers, rosy to lavender fading to silver with age. Small seeds are formed.

Plate 33
Koa
Acacia koa A. Gray
(endemic)

Koa is a chiefly high-altitude forest tree, preferring altitudes between 1,500 and 6,000 feet, although it will grow at lower altitudes, almost to sea level. When it grows singly, it has wide-spreading branches beginning low on the trunk and reaches heights of fifty feet. When growing closely together, as these trees do in the high forests, under ideal conditions (which includes the higher altitudes), the trunks are often tall and straight for a height of sixty feet before any branches begin; some trees of that height have diameters as great as ten feet. The bark is light gray, and smooth when the tree is young, but deeply furrowed longitudinally in mature trees. Upon germination of the seed and for a short time thereafter, the first true leaves are light green, finely divided, consisting of five to seven pairs of pinnae, each pinna with twelve to twenty-four pairs of leaflets. These are then replaced, as the tree grows, by dark green, smooth, stiff, crescent-shaped so-called leaves, which are actually broad, flattened petioles (leaf stems); these are called phyllodes and function as leaves. The trees bloom in late winter to early spring; the flowers appear as axillary clusters of small cream-colored "balls." Pods about three inches long follow; they are flat and contain small, flattened brown seeds. The wood resembles mahogany and, when polished, is a beautiful brownish red, through which the wavy lines of the grain show.

PLATE 34
Koai'a or *Koai'e*
Acacia koaia Hillebr.
(endemic)

Koai'a is a native Hawaiian tree that somewhat resembles its close relative, *Acacia koa,* but it grows to a height of only twenty to twenty-five feet; has a gnarly trunk; and its "leaves," actually phyllodes, are narrower and sickle-shaped, from five to six inches long and half an inch wide. The flowers form globose heads; the pods are narrow and curved. The wood is very hard.

Plate 35
Koali ʻawa or *Koali ʻawahia* (morning glory)
Ipomoea indica (J. Burm.) Merr.
(indigenous)

Koali ʻawa is a common native Hawaiian morning glory that grows almost everywhere except in dense forests, from sea level to an altitude of 2,000 feet. It has heart-shaped leaves and milky sap. The flowers are a delicate bluish violet when they open in the morning, gradually turn to pink through the day, and close at night. The vine drapes itself over grass and shrubs in fields, bypaths, and roads. In some localities the vines are covered with tiny silvery hairs and in other places the vines are hairless.

Plate 36
Kōlea
Myrsine spp.
(endemic)

Kōlea are native Hawaiian shrubs or trees, with narrow to oval leaves that are more or less crowded at the tips of branches. The *liko* (new flush of leaves) is usually quite attractive, being magenta to red. Flowers are small and usually occur in clusters on short lateral "spurs" usually below the attachment of leaves to the stem. Fruits are numerous; they are small, globose, yellow to black or red, and have one seed. The sap in the wood is red.

Plate 37
Kolokolo kuahiwi or *Kolekole lehua*
Lysimachia spp.
(endemic)

Kolokolo kuahiwi are small spreading or erect native Hawaiian shrubs or perennial herbs that belong to the primrose family. The branches are densely covered with leaves that are either narrow and pointed or broad and elliptical. The flowers are purplish or pinkish (rarely yellowish), generally solitary on slender stalks in leaf axils (in terminal clusters in one species). The capsules are globose or ovoid and open from the top into five to ten valves; seeds are numerous.

PLATE 38
Kolomona or *Kalamona*
Senna gaudichaudii (Hook. and Arnott) H. Irwin & Barneby
(indigenous)

Kolomona is a native Hawaiian shrub belonging to the pea family. It has wide-arching branches; the young shoots are greenish and silky-hairy. The leaves, which are finely hairy, consist of four to six pairs of elliptic-oblong leaflets. Flowers are greenish yellow and are borne in small axillary clusters. The pods are brown, minutely hairy, and up to six inches long and 1/2 inch wide. It is generally found in the arid lowlands and along the coast on the six main islands.

PLATE 39
Ko'oko'olau
Bidens spp.
(endemic)

Ko'oko'olau are primarily perennial herbs or shrubs with opposite leaves, simple and ovate or compound with three to seven ovate leaflets, varying in size from species to species, bearing few to many hairs. They are short- to long-stemmed. Flower heads may be clustered or single on long stems at branch tips; flowers have ray florets varying from white to yellow. Fruits are black or dark brown, narrow, short or long, and may or may not have barbed bristles.

PLATE 40
Kou
Cordia subcordata Lam.
(Polynesian-introduced)

Kou is an evergreen tree that reaches a height of about thirty feet, has a dense, wide-spreading crown, and a pale gray grooved and flaky bark. Its wood is beautifully grained, with wavy dark and light lines and bands. The leaves are about four inches long, with petioles from 1/2 to three inches long; they are ovate, long-pointed at the tip, and more or less rounded at the base. The blade is smooth and entire or wavy edged. Flowers are an inch long and at the mouth have a diameter of one to two inches. They are borne in short-stalked terminal and lateral clusters. They are orange and scentless. The calyx is three- to five-toothed, and the corolla is five- to seven-lobed. The fruit, which is enclosed in the enlarged calyx, is globose and about an inch long. When young it is green and when mature dry and grayish brown; it contains four seeds.

Plate 41
Kukui (candlenut tree)
Aleurites moluccana
(Polynesian-introduced)

Kukui is a large tree, tall and wide-spreading; it is found chiefly in the woods of the lower mountain zone. It is often concentrated in the ravinelike hanging valleys or the walls and floors of large valleys. Kukui trees are easily identified from a distance by their pale green foliage; the leaves are covered with a whitish "bloom," actually hairs. In shape, the leaves range from angularly pointed or lobed, somewhat like maple leaves, to narrow ovate; they may be as long as eight inches and have long petioles. Young branches are smooth, but a thick rough bark forms on the trunks of older trees. Underneath this outer bark lies the inner bark, which is brightly tinted with a reddish brown sap. The larger lateral roots that lie near or on the surface of the ground also contain this pigment. The tree also has a deep tap root. The flowers are small and whitish; they are borne in large clusters at the end of the branches. The fruit is globose, about two inches in diameter; it is borne on a long stem. Occasionally (in recent years, much more frequently), "twin" fruits are present. The fruit consists of a thick outer nonfibrous fleshy husk and a nut surrounded by a parchmentlike sheath. There is usually a single nut in a fruit, but in the twin fruits there are two, with one of the nuts sometimes smaller than the other. The nut, about as large as a walnut, is slightly domed on one side and flattened on the other; one end is blunt, the other has a slight projection—a "nose." When mature, the shell of the nut is black and shallowly furrowed; it is bony in texture and very hard. The kernels are white and adhere tightly to the shell. The kernels are very oily, yielding, on extraction, as much as fifty percent pure oil.

PLATE 42
Kūpaoa or *Na'ena'e*
Dubautia spp.
(endemic)

Kūpaoa are native Hawaiian resinous shrubs or small trees belonging to the sunflower family. They have narrow or broad leaves and small yellow, white, orange, or purple flower heads borne in large flat, cone-shaped, or pyramid-shaped clusters. The plants are often highly scented. The leaves are opposite, or alternate, or in threes (rarely in fours); they are quite variable in shape, from 1/2 to ten inches long, more or less toothed, usually stemless and clasping. Species of *kūpaoa* can be found in a wide range of habitats: alpine deserts; dry scrub; dry, mesic, and wet forests; and bogs.

PLATE 43
Lama
Diospyros spp.
(endemic)

Lama are native Hawaiian forest trees that belong to the ebony family; they are among the dominant tree species in dry forests at lower elevations. Their wood is hard and red-brown in color. The leaves are oval or oblong with pointed tips; they are from one to five inches long. They are reddish when young but leathery and dark green when mature. Male and female flowers are borne on separate trees; they are small and occur in the axils of the leaves. The fruits are small, ovoid, rather dry; they are about 3/4 inch long. The fruit turns from green to yellow to bright red. When ripe it is pleasantly sweet.

PLATE 44

Laua'e (*maile*-scented fern)

Phymatosorus scolopendria (N. L. Burm.) Pic.-Ser.

Laua'e creeps over the ground; it has broad, flat, oblong, shiny, dark green fronds, which are usually deeply lobed. The leaves arise singly, at intervals, on short or long stems four inches to three feet long, from brown-black, scaly, creeping rootstocks. Two types of leaves occur: vegetative and spore-bearing; the former type has broader lobes than the latter. The upper surface of spore-bearing fronds is dotted with elevations resulting from one or more rows, on each side of the prominent midrib, of rounded hollows, or sori, which are filled with uncovered sporangia on the lower surface.

Laukahi kuahiwi (plantain)

Plantago spp.

(endemic)

Laukahi kuahiwi are native Hawaiian members of the plantain family that grow at medium to high altitudes. Some forms are stemless, forming a rosette near the ground, but others are branched and reach heights of six feet. The leaves vary greatly in size from 1-1/2 inches long and 1/2 inch wide to twelve inches long and two inches wide. They are somewhat spatulate in shape, are fairly thick, and are smooth on both surfaces or covered by brown or tan wool beneath. They are slightly grooved with longitudinal veins. Tiny flowers are borne in cylindrical heads at the ends of slender stalks four to five inches high. A small capsule forms from each flower. These native plantains should not be confused with the broad-leaved plantain (*Plantago major* L.), which Hawaiians substituted for their own *laukahi kuahiwi* after *P. major* was introduced in postcontact times.

Lehua papa
Metrosideros rugosa A. Gray
(endemic)

Lehua papa is a native Hawaiian shrub or (rarely) small tree that grows in the high forests of O'ahu; it is distinguished by its rounded, leathery, and deeply grooved leaves that are densely woolly on the lower surface. The *liko* (new flush of leaves) is also densely woolly. The small flower clusters are solitary or in pairs in the axils of the youngest leaves. The flowers are deep red; the sepals are densely hairy, as are the branches of the inflorescence.

Plate 45
Limu (marine and freshwater algae)

The edible freshwater and marine algae of ancient Hawai'i are so numerous and variable as to locality, season, and regional nomenclature that it is impossible to name and describe them all here. One that was especially popular was *limu kohu (Asparagopsis taxiformis),* a soft, succulent seaweed with small; narrow; cylindrical; tan, pink, or dark red filaments that give this *limu* a tufted appearance, shown at 4 o'clock in Plate 45.

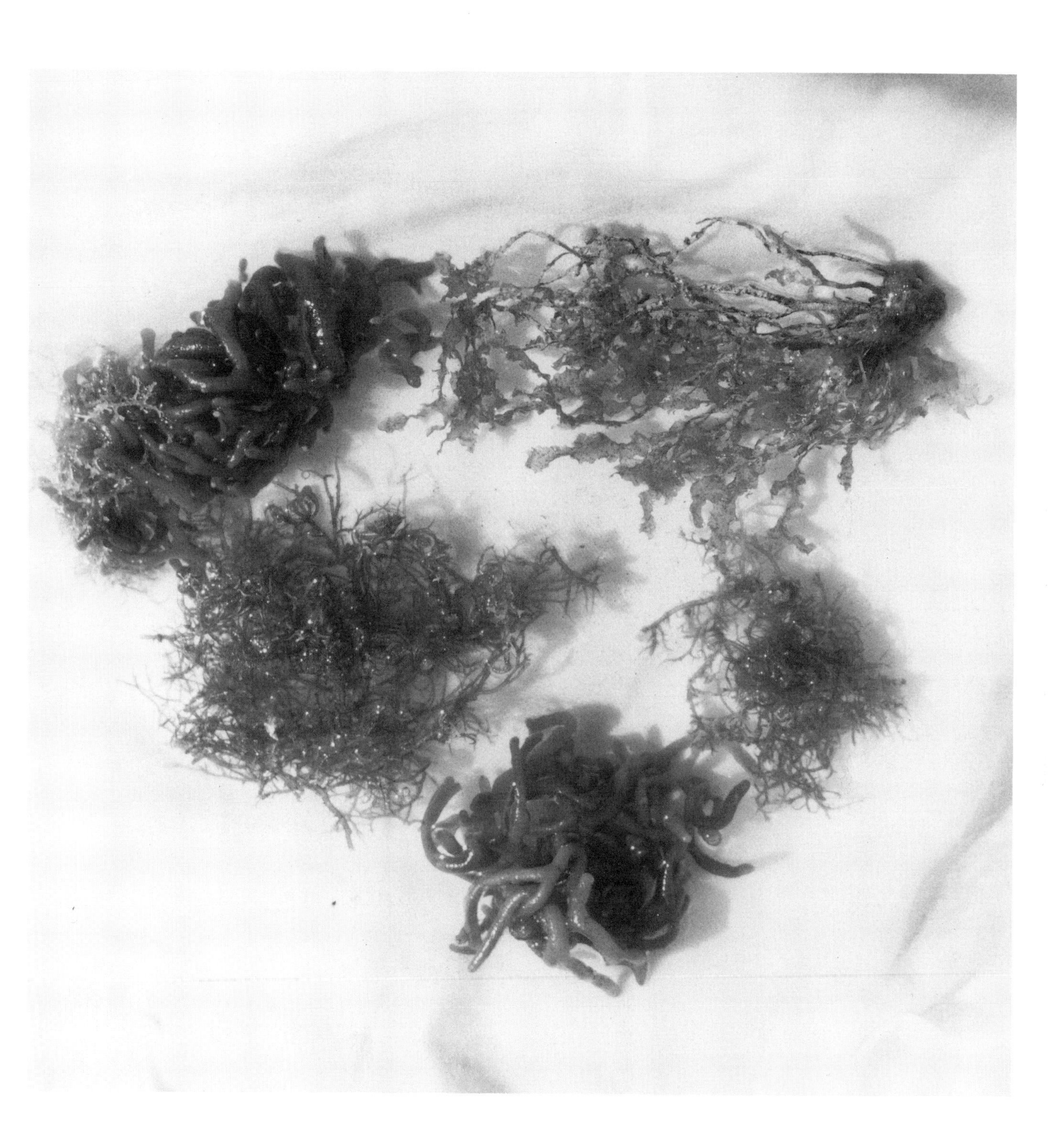

Plate 46
Loulu (fan palm)
Pritchardia spp.
(endemic)

Loulu is a name applied to several native Hawaiian fan palms. The leaves are borne in a cluster at the top of a ringed trunk, the rings being the scars of once-attached leaf stems. The leaves are very large, most are rigid, ordinarily broadly wedge-shaped and shallowly cut with numerous bifid segments (twice-divided clefts). The underside of the leaves, the strong spineless leaf stems, and the flower stems may have a waxy or feltlike covering. The numerous flowers occur in clusters; they are perfect, having a three-toothed calyx and a tubular corolla with three segments, the segments falling off upon the opening of the flower. There are six stamens, and a three-lobed and three-celled ovary, one cell only of which produces a globose or ovoid fruit that is a drupe. The fruit color ranges from green to yellow to brown or black at maturity. It is externally smooth, with a fleshy or fibrous layer covering a thin woody shell, which, in turn, covers a hard seed.

PLATE 47
Ma'aloa or *'Oloa* or *Ma'oloa*
Neraudia melastomifolia Gaud.
(endemic)

Ma'aloa is a spreading, climbing, or erect low native shrub with watery or milky sap, belonging to the nettle family. The leaves are alternate, densely minutely hairy to nearly smooth, ovate, elliptic or lanceolate, variously toothed or undulate or entire, and one to seven inches long and 3/4 inch to three inches wide, with three prominent veins. *Ma'aloa* is dioecious (male and female flowers are borne on separate plants). The inconspicuous flowers of either sex are sessile or very short-stalked and are borne in small axillary clusters. The red, juicy fruits, 3/16 to 1/8 inch long, enclose a seed 1/16 inch long.

Plate 48
Mai'a (banana)
Musa sp.
(Polynesian-introduced)

Banana plants are herbaceous perennials that assume treelike proportions, some varieties reaching heights of twenty-five to thirty feet, and "stem" diameters of six or more inches. What appears to be a stalk or trunk is made up of layers of spirally arranged, overlapping fibrous leaf bases. These grow from a great bulbous underground "base," which botanically is actually an underground stem or corm. It is because of the nonwoody nature of the so-called stem or trunk that bananas blow over so easily in windy areas. The petioles or leaf stems, continuations of the leaf bases that form the false trunk, are thick and may be long or short, depending on the variety. There is a groove or furrow in which the young furled leaf lies protected until it emerges and unfurls, with a "flag" at its tip. The leaves are very large, forming a rough rectangle as long as eight to twelve feet, with a width of two or more feet. There is a thick midrib. The blade itself is somewhat thick and smooth and more or less shiny on the upper surface; on the underside, it has a "frosted" appearance due to the presence of a bloom (a waxy layer). After its unfurling, the blade is entire but soon splits along the parallel side veins, often to such an extent, as in windy areas, that the leaf is "feathered." The flowering stalk grows up through the center of the trunk from the apex of the underground stem to issue from the top (the center of the leaf cluster) in the form of a large bud (an exception is the *hāpai* banana), when the plant is nine or ten months old. In all Hawaiian varieties that have an emergent flower stalk, this begins to turn downward so that the resultant inflorescence and fruit bunch are pendant. All but one Hawaiian variety produce a single inflorescence stalk; the exception is the variety *māhoe* (twins) in which the single stalk divides, with each branch then bearing a bunch of fruit. The flowers are spirally arranged on the flower stalk, in consecutive clusters, extending from the base to the apex. One or two clusters of flowers open at a time. Each developing cluster is completely protected by a thick, close-fitting, leaflike bract of various shapes and colors depending upon the variety. The outer surface of the bract may be purple, claret, or reddish brown, and is powdered with a frostlike bloom similar to that on the underside of the leaf blade. The inner surface of the bract may be red, pink, purple, violet,

yellow, orange, brown, or a blend of several of these colors. The outside surface may be shallowly ridged, but the inner surface is usually smooth and "glossy." In the early stages of flower development, the bracts are tight-clasping, so that the inflorescence "bud" forms a budlike structure. As the flowers mature, the subtending bract becomes raised, rolls back, and eventually loosens and falls off. The female (pistillate) flowers are the first to develop. After various numbers of the two-row clusters of female flowers are initiated and their ovaries begin to form fruits, a series of "neutral" flowers, with undeveloped stamens and pistils, form along the elongating flower fruit stalk. When several rows of these have developed, the male (staminate) flowers are initiated at what is the tip of the stalk. These last-formed flowers usually remain enclosed beneath the tightly appressed bracts and may still be in this condition at the time of fruit harvest. Although it might seem advantageous, from the standpoint of nutrient accumulation in the developing fruit, to remove that part of the flower/fruit stalk bearing the neutral and staminate flowers, there is some thought that hormones from these "zones" are beneficial to the developing fruit. The ovaries of the individual pistillate flowers develop into individual fruits (botanically called berries); the fruits are more or less cylindrical, with various skin and flesh colors, depending upon the variety. The individual fruit are commonly called "fingers"; the two-row groups of fruit, a "hand"; and the entire collection of fruit, a bunch. The fruit curve upward on the pendant stalk of the Hawaiian varieties. The fruit develop in sixty to eighty days after the inflorescence pushes out of the center of leaves. Each trunk bears one bunch of bananas except the previously mentioned variety *māhoe,* which bears two bunches on a divided fruit stalk, and the *hāpai* banana, the stalk of which does not elongate, and the stem of which splits to reveal a few bunches of small bananas.

PLATE 49

Maiapilo or *Pilo* or *Pua pilo*
Capparis sandwichiana DC
(endemic)

Maiapilo is a native Hawaiian member of the caper family. It is a straggling shrub two to three feet high with spreading, horizontal, yellowish branches. It grows chiefly on beaches or on lava flows in the lowlands. The leaves are oblong with rounded ends; they are about two inches long and an inch wide. They have petioles about 1/2 inch long; there are no stipules. The delicate, beautiful, fragrant flowers consist of four wide, white petals two inches long surrounding a mass of protruding, white, yellow-tipped stamens. The flowers are borne singly, on stalks two inches long, in the leaf axils; they are nocturnal and soon wilt during the heat of the day.

PLATE 50
Maile
Alyxia oliviformis Gaud.
(endemic)

Maile is a native Hawaiian shrubby vine, straggling on the ground or twining on nearby shrubs or trees in the native forests of the lower and middle mountain regions. The leaves are very variable in shape and size. The most common are oval, pointed, and shiny, with short stems. They grow in pairs, threes, or fours and are one to three inches long. Three or four very small yellowish flowers, each just under 1/10 inch long and having four or five short lobes, are borne in the axils of the leaves. The fruit is an ovoid drupe 1/2 inch or longer, often with two or three in a row, like a string of beads. Several forms of maile are recognized. One form has wider, blunt, oval leaves occurring in pairs; another form has narrower leaves in pairs, threes, or fours; and a third form has smaller, oblong to round leaves in threes and fours, and smaller fruits. The bark of the stems, and the leaves, are highly scented, when bruised, as a result of oxidation of the sap.

PLATE 51
Makaloa
Cyperus laevigatus L.
(indigenous)

Makaloa, native to Hawai'i and other tropical regions, is a perennial sedge found in or near fresh or salt water. Many long, erect stems, 0.12 inches wide or less and four to twenty-eight inches tall, arise from creeping rhizomes. Each stem is leafless or accompanied by a single short, linear leaf and is topped by a small laterally appearing inflorescence consisting of one to several stemless spikelets 1/8 to 3/4 inch long. Each inflorescence is accompanied by two bracts, the upper one stiff, erect, and appearing as a continuation of the stem.

PLATE 52
Māmaki or *Māmake*
Pipturus spp.
(endemic)

Māmaki, native members of the nettle family, are shrubs or small to medium-sized trees, up to thirty feet tall. They are usually found on the outskirts of forests and in clearings, generally at altitudes between 1,500 and 4,000 feet. They have variously toothed leaves about four inches long that are usually white beneath. The leaves vary in texture from thin and flaccid to stiff and leathery; the veins on the lower surface are green, reddish, or purplish. The bark is smooth and light brown and is matted with gray, woolly hairs on young branches. The fruit are dry, brown achenes embedded in a white, tasteless, and somewhat fleshy receptacle.

Plate 53
Māmane or *Mamani*
Sophora chrysophylla (Salisb.) Seem.
(endemic)

Māmane, a native member of the pea family, is found on all the main islands except Ni'ihau and Kaho'olawe at altitudes between about 1,000 and 9,000 feet. It has different forms: erect trees twenty to forty feet high; sprawling trees near the ground; and erect shrubs. The leaves are five to six inches long and have thirteen to twenty-one oblong leaflets that are almost paired or distinctly alternate, 3/4 to 1-1/2 inches long and 1/2 inch wide; they are smooth or bear gray or yellowish down on the underside. The young growth is also downy. The flowers, an inch long and yellow, are borne in small clusters at the tips of branches and in the leaf axils. The pods are quadrangular, four-winged, four to six inches long and 1/3 inch wide. They are more or less constricted between the four to eight yellow, oval seeds.

PLATE 54
Ma'o (Hawaiian cotton)
Gossypium tomentosum Nutt. ex Seem.
(endemic)

This native cotton is a low shrub growing primarily in arid coastal areas. Stems and leaves have soft white hairs. Leaves are wider than long and usually three-lobed. Flowers are usually solitary with bright yellow petals. Capsules are woody and three-celled. There are two to four small seeds per cell, each seed covered with short reddish brown lint.

Plate 55
Milo (portia tree)
Thespesia populnea (L.) Sol. ex Corrêa
(indigenous)

Milo is a tree that rarely reaches a height of forty feet. The branches ordinarily spread widely, more or less horizontally. The trunk attains a maximum diameter of two feet. The bark is thick and corrugated; the twigs are scaly. The wood is reddish brown and has a beautiful grain. The leaves are heart-shaped, three to five inches in width, and are a bright and glossy green. The flowers are bell-shaped, from two to three inches in diameter. The petals are a pale yellow with purplish base (the center of the flower shows this color). During the day the flowers wither into a purplish pink twist. The fruit (the seed cases) are about an inch in diameter, globose, slightly flattened on each end; they are five-celled, and each cell contains a woolly seed. When mature the capsules, which do not open, are brownish and somewhat woody.

PLATE 56
Moa
Psilotum nudum (L.) P. Beauv.
(indigenous)

Moa is a slender, more or less erect, shrubby or tufted perennial up to about one foot tall, with many successively two-forked branches that are ridged longitudinally. Tiny, pointed, scalelike "leaves" are scattered fairly uniformly over the branches. The stems themselves are green and function as leaves. Among the upper scales are borne three-lobed yellow cases filled with countless minute spores. A slender underground stem functions as a root; it contains a tiny fungus that serves to absorb food materials and water. *Moa* grows in moist to dry localities, on soil, among rocks, and on trees. It is widely distributed because of its propagation through spores.

Plate 57
Mokihana
Pelea anisata H. Mann
(endemic)

Mokihana, a member of the rue family, is a native tree that grows only on the island of Kaua'i. The leaves are simple and opposite; their blades are three to seven inches long and one to two inches wide. Their tips are blunt or have a notch. The flowers are very small, with one to three on a short stem. The fruit is a leathery, cube-shaped capsule about half an inch in diameter. It opens into four parts, each part with two black, shiny seeds. All parts of this plant are anise-scented.

PLATE 58
Naio (bastard sandalwood)
Myoporum sandwicense A. Gray
(indigenous)

Naio is a small to large native Hawaiian tree that grows at altitudes near sea level up to about 7,000 feet, being common on the slopes of high mountains such as Mauna Kea, Mauna Loa, and Haleakalā. It has dark gray, grooved bark. The leaves are alternate, narrow-oblong, pointed, rather thick, entire or toothed, smooth or hairy, two to six inches long, with translucent spots; they are grouped at the ends of branches. The flowers are small, pink or white, commonly five-parted (sometimes four to nine parted), with a superior ovary. They form clusters of five to eight along the twigs. The fruit is a globose drupe that is white, pinkish, or purplish; somewhat juicy; and about a quarter of an inch in diameter. It contains four to twelve cells joined together in a hard, ribbed structure. The heartwood of the trunk and branches is scented much like that of *'iliahi,* the "true" sandalwood.

PLATE 59
Nānū or *Nā'ū* (native gardenia)
Gardenia spp.
(endemic)

Nānū are three native Hawaiian gardenias, all of which grow in the forest. Each has white, tubular, fragrant flowers. *Gardenia brighamii* H. Mann is a shrub or small tree six to eighteen feet high with ovate leaves to four inches long. *Gardenia mannii* St. John & Kuykendall and *G. remyi* H. Mann are trees reaching heights of thirty to forty-five feet. Their leaves are elliptic to inverted ovate to oblong and are four to nine inches long. They differ in that the calyx spurs of *G. mannii* are narrow and are somewhat curved, while those of *G. remyi* are wider and straight. The fruit of *G. brighamii* is globose and an inch in diameter; that of *G. remyi* is pear-shaped, four- or five-angled, and up to two inches in diameter. *G. brighamii* is officially listed as an endangered species by the U.S. Fish and Wildlife Service and is thus protected by Federal as well as State of Hawaii statutes.

Plate 60

Naupaka kahakai (beach *naupaka*)
Scaevola sericea Vahl
(indigenous)

This native Hawaiian plant, *naupaka kahakai,* is a smooth, spreading succulent shrub about three to ten feet high; the plant branches freely from its base. At the tip of the branches are tufts of fleshy, bright green convex leaves three to five inches long; these are inverted ovate, rounded, and notched at the tip. The leaves and inflorescence are more or less silky downy. Clusters of five to nine flowers, each about 3/4 inch long, are borne in the axils of the leaves; they are white, often streaked with purple and are fragrant. On emerging, the flowers split longitudinally, giving the appearance of half flowers. The berries are white, succulent, two-celled, and about 1/2 inch long.

Plate 61
Naupaka kuahiwi (mountain *naupaka*)
Scaevola spp.
(endemic)

These native shrubs or small trees, 1-1/2 to 12 feet high, have smooth to hairy young branches with thin, smooth to densely hairy, oval or narrow, light green leaves. More than half a dozen leaves grow in a cluster. The flowers of these *naupaka* are smaller than that of the beach *naupaka* but like the latter look like a half flower. Flowers are either white, yellow, or lavender; the corolla has five winged lobes and a tube split halfway or more down one side. There are five short stamens forming a ring around the style, which protrudes beyond them. The black (or rarely white) fruit is drupaceous and contains two stones.

PLATE 62
Nehe
Lipochaeta spp.
(endemic)

Nehe are native to Hawai'i; they belong to the sunflower family. These plants are prostrate to erect perennial or rarely annual subshrubs or herbs, growing from sealevel to about 6,000 feet in elevation. They have numerous leaves that are opposite, ovate to spatula-shaped with entire, toothed, or lobed margins, and hairy or scabrous on both sides. The blades, about an inch long, taper into short or long stems. The flower heads range from solitary to clustered and have four to fifteen yellow ray florets surrounding yellow disk florets. The fruits have a short bristly pappus.

PLATE 63
Niu (coconut)
Cocos nucifera L.
(Polynesian-introduced)

Niu, a palm, has a slender, more or less curved and leaning, graceful, ringed trunk that is thickest at the base. It can live for a hundred or more years, reaching a height of 100 feet. The leaves (fronds), in a spreading cluster at the top of the trunk, are nearly flat and six to eighteen feet long. They have a long, smooth, fibrous stem, the base of which spreads to encircle the trunk. Leaves have numerous narrow leaflets with conspicuous midribs; they are from one to three feet long. Trees may bear flowers and fruit when six years old, and produce up to forty or more fruits a year. The flowers are borne in simple-branched clusters among the leaves. The male flowers are small and arranged along the long branches of the inflorescence; the few female flowers occur at the base of each branch. The fruit, a drupe or nut, ripens in nine to ten months. It is more or less triangular in cross section, varies from nearly round to oblong, and is eight to twelve inches long. A thick fibrous husk, with a smooth skin, surrounds a globose, hard, thin-shelled nut having three pores at one end; the one above the single embryo is soft. Lining the inside of the shell of the ripe nut is a layer of white, solid, edible, sweet pulp that is commonly called "flesh" or "meat." This pulp surrounds a hollow containing sweet water.

Plate 64
Noni (Indian mulberry)
Morinda citrifolia
(Polynesian-introduced)

Noni is a small evergreen tree growing in open lowlands, often on lava, and at edges of forests; it rarely attains a height of twenty feet. It has coarse branches, angular in cross section; branchlets are thick and conspicuously marked with leaf scars. The leaves are opposite, large (eight or more inches long), ovate, thick, deeply veined, short-stemmed, dark green, and shiny. Flowers develop in small, globose, fused "heads" to form an inflorescence about an inch in diameter. Each head bears many flower buds, of which only one or two open at a time. Each flower consists of a white, tubular, five- to seven-lobed corolla about 1/3 inch long. Flowers are closely packed and appear in various stages of development in an expanding inflorescence—near the apex there are still green buds while older flowers, near the base, have opened and closed. Before the last flower at the apex of the flower cluster has bloomed, small pale yellowish-green fruitlets have replaced the flowers. When the fruit is mature, it is several inches long, usually elongated in shape, and "warty" in appearance because of the uneven growth of individual fruitlets. The surface is marked with the more or less faint pentagonal or hexagonal outlines of individual fruitlets. "Pits" within each of these "outlines" are the circular scars of the corolla. When mature, the skin is whitish, and the flesh is whitish yellow and insipid. When ripe or overripe, the skin becomes translucent and the flesh soft and very foul smelling. The seeds, sometimes called kernels, are oblong-triangular and reddish brown. A bladderlike air sac is attached to each seed. This gives buoyancy to a detached seed; it is believed that the wide distribution of this plant may, in part, be the result of this flotation mechanism.

Plate 65
Nuku ʻiʻiwi or *kā ʻiʻiwi*
Strongylodon ruber Vogel
(endemic)

Nuku ʻiʻiwi, a legume, is a native Hawaiian woody vine growing in the forests on all the main islands except Lanaʻi. The leaves consist of three leaflets from 2-1/2 to five inches long and two to three inches wide. The flowers are red and are shaped like narrow beaks, hanging in narrow clusters. The pods are thick, flattened, smooth, about four inches long and two inches wide. There are one or two round seeds, which are black; the scar is semicircular.

Plate 66
'Ōhā or *'Ōhā wai* or *Hāhā*
Clermontia spp.
(endemic)

'Ōhā are plants native to Hawai'i and belong to the bellflower family; they are forest trees or shrubs. Clermontias are woody plants, either shrubs or small trees, with stems branching often in a candelabra-like fashion. The leaves are alternate, simple, or toothed; there are no stipules. The flowers usually occur in twos or threes and are perfect; the calyx and corolla are five-parted; the calyx of some of the species resembles in size and color the corolla, causing the flowers to appear "double." Species of *Clermontia* have flowers that are long, slit down the back, and curved like the beak of native nectar-feeding birds that suck honey from them. The flowers of most species are showy, ranging from green to white to purple.

PLATE 67
'Ohai
Sesbania tomentosa Hook. and Arnott
(endemic)

'Ohai, a native coastal or low-elevation Hawaiian legume, is a low-growing, spreading prostrate shrub, or a small tree to about eighteen feet tall, with silky wool-covered branches and leaves, and narrow, red to salmon-colored (rarely yellow) flowers that are an inch long. The silvery green leaves are about four inches long and consist of about nine to nineteen pairs of leaflets. The flowers are borne in small clusters in the axils of the younger leaves. The pods are narrow and from five to seven inches long and enclose six to twenty-seven small, olive-green to brown seeds.

PLATE 68

'Ohe (bamboo)

Schizostachyum glaucifolium (Rupr.) Munro

(Polynesian-introduced)

The "Hawaiian" bamboo, *'ohe,* is a giant grass with erect stems growing from vigorous underground stems. The hollow internodes are long, with thin walls; nodes, forming diaphragms, occur at more or less regular intervals. One to many branches occur at the nodes. The "walls" of the internode are fibrous and hard in older sections of the stem. The plants form rather open clumps, nine to fifteen feet high. The leaves are long, flat, and pointed, with a stemlike base joined with the leaf sheath, with many longitudinal veins; the bases are uneven-sided. The leaves are covered with a whitish bloom. Seen in flower often, the inflorescence consisting of many slender branches from a node; the spikelets are in tightly packed clusters.

PLATE 69
'Ōhelo
Vaccinium spp.
(endemic)

'Ōhelo, natives of Hawai'i, belong to the heath family; they are shrubs about two feet high (six feet tall in one species), with leathery, oblong to nearly circular leaves, about an inch long. The margins are finely serrated. Leaves often are bluish green or grayish green. The flowers develop singly in leaf axils and are usually red (green in one species). The fruits, about 1/3 inch in diameter, are either globose, flattened, or somewhat elongated; they are either red, orange, or yellow and contain a considerable number of small, flattened seeds. The shape and color of the berry vary with the species found in different island habitats. The berry is edible and highly prized because of its flavor.

Plate 70

ʻŌhiʻa ʻai (mountain apple or Malay apple)
Syzygium malaccense (L.) Merr. & Perry
(Polynesian-introduced)

ʻŌhiʻa ʻai is a fast-growing tree found in shady valleys at altitudes up to about 1,500 feet. It is often seen in groves, where it reaches a height up to fifty feet. When it grows solitarily, it tends to have a shrubby appearance. The branches have a smooth, mottled, gray bark. In spring, cerise, pompomlike flowers on short stems appear on the trunk and branches. These flowers, fallen to the ground to form a cerise-colored carpet, are a beautiful sight. The fruit is about as large as a medium-sized tomato, somewhat bell-shaped. It has a thin, deep crimson skin, with pure white, crisp flesh. It is slightly sweet but does not have any distinct taste. It is, however, very refreshing because it is so juicy. It contains one large, rounded seed. This red-fruited mountain apple was Polynesian-introduced. There is also a white-flowered form with white fruit that is seedless.

PLATE 71
'Ōhi'a hā or *Hā*
Syzygium sandwicensis (A. Gray) Nied.
(endemic)

Hā is a native forest shrub or tree up to seventy-five feet tall, with dark green, inverted ovate leaves, two to four inches long. The small white flowers with many white to pinkish stamens are clustered in the axils of the upper leaves. The fruit is globose, about 1/3 inch in diameter; they are red and one- or two-seeded, with little pulp that is edible. *Hā* is related to the mountain apple, *'ōhi'a 'ai.* It is also related to *'ōhi'a lehua,* from which it is easily distinguished by its smooth, reddish brown bark and fleshy fruit.

Plate 72
ʻŌhiʻa lehua or *Lehua* or *ʻŌhiʻa*
Metrosideros polymorpha Gaud. and *Metrosideros macropus* Hook. & Arnott
(endemic)

ʻŌhiʻa lehua are native Hawaiian plants that are extremely variable, ranging from tall trees to low shrubs; leaves vary from round to narrow, with tips that are blunt or pointed and with surfaces that are smooth or woolly. In texture they may be thin and soft or leathery and stiff. Young leaves usually are reddish or chartreuse or pink; older leaves are dull or shiny green. *Metrosideros polymorpha* flowers are usually a dark or lighter red or orange-red; less often they are salmon, pink, yellow, or white; *M. macropus* flowers are yellow, rarely red. The color is in the numerous stamens, which are from half an inch to an inch in length. The flowers appear as tufts at the ends of branches. The bark is rough and scaly and the wood is hard and dark red. The fruit is a capsule filled with many tiny, linear seeds.

Plate 73
'Ōlapa or *Lapalapa*
Cheirodendron spp.
(endemic)

'Ōlapa are native Hawaiian forest trees with opposite leaves. Each leaf is divided into three to five leaflets that are oval or oblong in one species, or three leaflets that are wider than long in another species. The numerous flowers are borne in umbels in the leaf axils. The fruits are juicy, deep purple when ripe, ovoid or subglobose, and about 1/4 inch in diameter. The plant is characterized by the fluttering of its leaves, even in the slightest breeze.

Plate 74
'Ōlena (turmeric)
Curcuma longa L.
(Polynesian-introduced)

'Ōlena belongs to the ginger family. The plant is stemless, with several leaves rising in a cluster, twenty inches or higher, directly from underground stems (rhizomes). The "flesh" of the rhizome varies from a canary-yellow when young to a deep orange or mustard when older, in contrast to the white flesh of rhizomes of other gingers. The petioles, eight or more inches long, overlap to form a false stem. The leaves are thin, light green, and eight inches long and three inches wide. The plant lies dormant during the winter months, with the leaves dying back in the fall and new leaves appearing in the spring. The inflorescence is cylindrical and about three inches long; it appears in the center of the leaves. It consists of large, pale green, pouchlike, curved bracts, each subtending three pale yellow flowers, except in the upper part where the bracts are white or pinkish and without flowers. During some years very few if any seeds form; in others, many seeds are to be found. The fruit is a capsule that bursts open to reveal rows of bright red seeds.

PLATE 75
Olonā
Touchardia latifolia Gaud.
(endemic)

Olonā, a native member of the nettle family, grows in deep, moist ravines (gulches) in the mountains on the six main islands in Hawai'i, and is characterized by its prominent stipules, which are two to three inches long. It is a sparingly branched shrub, from four to eight feet high, sometimes growing rather prostrate. The leaves are large, from nine to sixteen inches long and five to nine inches wide, ovate with serrated (fine-toothed) margins, and green on both sides. Male flowers are borne in dense clusters 1/2 to one inch in diameter and three to five inches long. Female flowers are borne on smaller heads in shorter clusters. The fruit is roundish and mulberry-like.

Plate 76
Pa'iniu
Astelia spp.
(endemic)

Pa'iniu are native Hawaiian lilies that form a rosette growing either on the ground or on trees. They have a short stem or none. The leaves are long and narrow, 3/4 foot to four feet long and 1/2 to three inches wide. They are covered with silvery hairy scales and usually have three prominent veins. The inflorescence is a panicle and is shorter than the leaves. It is also covered with silvery hairy scales. The fowers are small, yellow or greenish, with the two sexes on different plants. The fruit is an orange berry with several seeds.

Pala
Marattia douglasii (Presl) Baker
(endemic)

Pala is a native Hawaiian fern with a short trunk and large, long-stemmed, much divided, dark green fronds up to eight feet long and five feet wide. The bases of the frond stems, which are hoof-shaped, thick, and starchy, are surrounded by two persistent, fleshy, black projections. Unlike *hāpu'u,* the spore cases of *pala* are not on the margin of the pinnae. *Pala* is found uncommonly in dense wet forests on the six main islands.

PLATE 77
Pala'ā (lace fern)
Sphenomeris chinensis (L.) Maxon
(indigenous)

The lace fern is one of the most common of the wild Hawaiian ferns. From an underground stem, light brown, smooth, shiny stems, one-half foot to a foot long, rise to support smooth, ovate, pointed fronds that are as long as the stem or longer. The fronds are subdivided three times; the smallest divisions are wedge-shaped. Each division bears on its underside, close to the edge, one or two sori with a covering only at the top.

Plate 78
Palapalai or *Palai*
Microlepia strigosa (Thunb.) K. Presl
(endemic)

This fern is similar to the lace fern in having three subdivisions of its fronds, but it has larger, "softer" appearing fronds commonly one to three feet long, and is hairy. The rhizomes are rather stout and covered with pale yellowish hairs. Spore cases are found at the tip of the veinlets of the last segments of the fronds, or above the first fork of a veinlet.

PLATE 79
Pāpala
Charpentiera spp.
(endemic)

Pāpala are forest trees native to Hawai'i. They have alternate, lanceolate, elliptic, ovate or obovate leaves that are membranaceous or leathery, often with a thickened margin. When young, the leaves are slightly to densely hairy. Older leaves are smooth. The numerous tiny flowers are borne in slender, smooth or minutely hairy compound panicles four to twenty-four inches long in the leaf axils. The wood is light and flammable. *Pāpala* are members of the amaranth family.

PLATE 80
Pia (Polynesian arrowroot)
Tacca leontopetaloides (L.) Kuntze
(Polynesian-introduced)

The leaves of *pia* are borne on finely but distinctly grooved petioles, one to three feet long; these arise directly from an underground tuber. This tuber is usually round, although it is sometimes somewhat flattened. In shape and skin color it somewhat resembles the white or Irish potato. The interior of the tuber is hard and white, and filled with fine-grain starch. When the plant is grown in loose soil, the tubers are found some distance from the base of the plant itself. The leaves are long and broad—one or more feet across—and are much divided, having somewhat the appearance of papaya leaves. The flower stalk is up to a yard high; at its top is a striking inflorescence consisting of three dozen or more small green and purplish flowers with short included stamens and styles. The flowers are subtended by six to twelve leafy bracts and numerous long threadlike bractlets; these latter are spreading and drooping, green and purplish, and four to nine inches long. By fall the leaves have wilted and by winter they have dried back completely; the tuber then lies dormant for three to four months. With the growth of new leaves additional tubers are formed.

Pi'a (wild yam)
Dioscorea pentaphylla
(Polynesian-introduced)

The cross section of the stem of *pi'a,* one of three Polynesian-introduced yams, is round; the leaves are palmate, with three to five lobes. This yam bears aerial tubers, spherical or horseshoe-shaped; these are scattered along the stem, in the axils of the leaf petioles. The underground tubers vary greatly in shape and size. Their exterior is covered with scattered, stiff, woody-spiny rootlets. The flesh of the tuber may be white, reddish, or brown. Male and female flowers are borne in separate inflorescences on the same plant; both types of flowers are small and inconspicuous. The male flowers are yellow and fragrant. A fruit forms from a fertilized female flower.

Plate 81
Pili (twisted beardgrass or tanglehead)
Heteropogon contortus (L.) P. Beauv. ex Roem. & Schult.
(indigenous)

Pili is a perennial grass found in open, dry, and sometimes rocky land. The plant forms tufts as high as three feet; it has leaf blades four to twelve inches long and about a quarter of an inch wide. The flowers are borne on narrow, crowded spikes that are up to four inches long. The spikelets overlap; each fertile one bears a conspicuous red-brown awn or bristle about four inches long—this is made crooked with two bends, hence its common name. This plant should be called simply *pili* because to call it *pili* grass, as is usual, is redundant. In years past, this grass was abundant on the dry ridges extending from St. Louis to Koko Head; with the establishment of housing subdivisions on these ridges, *pili* is not common on O'ahu. It can more commonly be found on other islands.

PLATE 82

Pōhinahina or *Kolokolo kahakai* (beach vitex or chaste tree)
Vitex rotundifolia L. fil.
(indigenous)

Pōhinahina is a native Hawaiian aromatic creeping beach shrub, with erect side stems usually four to twenty inches high; and downy, simple leaves, one to two inches long, round to inverted ovate. The flowers are blue and borne in clusters, one inch or less wide, at the tips of branches. The fruit is a reddish to black globose drupe about 1/4 inch in diameter.

PLATE 83
Pōhue or *Ipu* (gourd plant)
Lagenaria siceraria (Molina) Standl.
(Polynesian-introduced)

The gourd plant is a downy, annual, wide-spreading vine, with branched tendrils and round or heart-shaped leaves; these are five-angled or lobed, four to six inches in diameter. Flowers are night-blooming, white, occuring solitary in the leaf axils, and are about 1-1/2 inches long. Female flowers are shorter-stemmed than male flowers. The skin of the fruit is green in color when immature, and beige or tan when mature. The internal pulp is white; the flat seeds are light-colored and about 1-1/2 inches long. The thickness of the skin wall varies greatly with the variety; the rind becomes hard and woody and is long lasting. The various shapes of the gourd fruit are discussed in the chapter on food.

It is difficult to find the "native" gourds in Hawaii today. Varieties grown here today are of mainland or Mexican origin as are the fruit sold in shops, either as unprocessed fruit or made into musical instruments and other artifacts.

Plate 84

Pōhuehue (beach morning glory)

Ipomoea pes-caprae subsp. *brasiliensis* (L.) Ooststr.

(indigenous)

Pōhuehue is a vigorous vine creeping on sandy beaches above the high-water line. It has green stems several yards long and forms roots at the joints (nodes). The main root is long; one has been found that was twelve feet long and two inches thick. The leaves are thick and broad; they are inverted heart-shaped, or shaped like a goat's foot (notched at the tip). Sometimes they are broader than long, with the two lobes folded up along the midvein. The flowers are bell-shaped, a dusty pink to purple, blooming singly or a few in a cluster. The greatest number open in the morning. The fruit is a small, round capsule that contains four downy seeds.

Plate 85
Pōpolo (glossy nightshade)
Solanum americanum Mill.
(indigenous)

Pōpolo is a member of the nightshade family. It is herbaceous, from one to three feet high. It has small, thin ovate leaves, from one to four inches long, with more or less wavy edges. Three to ten white flowers, each about 1/4 inch wide, are borne in pendant, umbel-like structures. Ripe fruits are purplish black, juicy, edible berries about 3/8 inch in diameter.

Plate 86
Pua or *Olopua*
Nestegis sandwicensis (A. Gray) Degener, I. Degener, & L. Johnson
(endemic)

Pua is an evergreen tree belonging to the olive family; it may reach a height of sixty feet. It is found in low-altitude forests. The leaves are leathery; rather narrow, three to six inches long; and light green on their undersides. The flowers are perfect, pale yellow, and clustered at the leaf axils. The fruit is an ovoid dark blue drupe about 1/2 inch long. The wood is hard, heavy, strong, and dark brown with black streaks.

Plate 87

Pua kala (prickly poppy)

Argemone glauca (Nutt. ex Prain) Pope

(endemic)

Pua kala is a native Hawaiian poppy that grows in dry, rocky soil from sea coasts to an altitude of 6,000 feet. The plant, an annual, is attractive, with the silvery gray coloration of its leaves. It is about one to four feet high with coarse prickly leaves alternating along a moderately prickly stem that they clasp halfway around. They are wavy, lobed, and toothed. The plants bloom much of the year; the flowers, which are found at the tips of branches, have six petals spread open, forming a flat corolla three inches across. The petals are delicate and white. The center of the flower is filled with orange-tipped stamens that encircle a red-tipped pistil. The fruit consists of prickly capsules about an inch long; these contain seeds that are scattered when the capsule bursts open on maturity. These seeds can resist fire, so that new plants sprout and grow in recently burned-over ground.

PLATE 88

Pūkiawe or *'A'ali'i mahu* or *Kānehoa* or *Kāwa'u* or *Maiele*
Styphelia tameiameiae (Cham. & Schlechtend.) F. V. Muell.
(indigenous)

Pūkiawe is a shrub native to Hawai'i; it grows from three to nine feet high, occasionally higher. The plant develops numerous fine, brittle stems and branches. Leaves are numerous and very small, about 1/2 inch long and about 2/10 of an inch wide. They are alternate, stiff, leathery, very short-stemmed, and oblong with the tip blunt or pointed and the base blunt or wedge-shaped. The upper surface is smooth while the undersurface appears whitish from bloom. The plants bear little, inconspicuous flowers in the leaf axils. These are five-parted with the corolla a tube. The fruit is a globose, pink, red or white, rather dry drupe, about a quarter of an inch in diameter; the seeds are ovoid.

PLATE 89
'Uala (sweet potato)
Ipomoea batatas (L.) Lam.
(Polynesian-introduced)

Most varieties of *'uala* are vines, growing close to the ground; a few varieties are vineless and thus are bushy. Leaves vary greatly in shape, from heart-shaped to angled, and are partially or deeply indented (lobed). Although the blades of leaves are normally always green, the veins and leaf stems (petioles) may vary in color, ranging from green through pink to purple. Handy reported that sweet potato plants do not flower abundantly except when neglected. However, Chung reported that this plant "blooms profusely in Hawaii from November to April" (this latter is also my own observation). The flowers are pinkish lavender, with the corolla (petals) forming a tube below and spreading out widely above. The roots increase in size (diameter) to form large roots; most of these grow closely under the center of the plant. The swollen roots vary greatly in size and shape; they are elongated with a small diameter, or spindle-shaped, or almost spherical. The Hawaiians also made note of whether the surface was smooth or grooved. When grown in rocky soil or in the presence of roots of neighboring trees, the shape varies from these norms, becoming considerably distorted; this would not have happened in old Hawai'i, where the farmer usually planted sweet potatoes in well-prepared beds. The color of the skin, that of the cortex or flesh immediately beneath the skin, and that of the flesh itself ranges from an almost pure white or light cream through pink, to red, purple, or a purplish red. Sometimes each of these three tissues has different colors in the same swollen root.

Plate 90
'Uhaloa
Waltheria indica L.
(indigenous)

'Uhaloa is a small shrubby, down-covered plant with blunt-oval, toothed velvety leaves from one to 3-1/2 inches long, conspicuously veined. Flowers are small, yellow, and five-parted, densely clustered at leaf axils. The fruiting capsules are tiny, two-valved, and two-seeded.

Plate 91
Uhi (yam)
Dioscorea alata L.
(Polynesian-introduced)

Uhi, like most yams, is a vine. The stem of this particular yam is angular (square in cross section), with longitudinal ribbonlike wings; these wings are greenish or reddish, depending upon the variety. The stem has a tendency to become twisted as it elongates. The leaves, arranged oppositely on the stem, are light green, heart-shaped, and from three to six inches long. There are seven to eleven veins, joined at the base and apex of the leaf. Uhi does not bear aerial tubers in the axils of the leaves as do the other two Hawaiian yams. The plant produces tubers that are variously shaped depending upon the variety: elongated with much the appearance of a very large carrot or daikon (white radish), to flat and lobed; in the latter case the tuber is sometimes deeply lobed and has the appearance of a human hand. Tubers grow deeply and may attain a great size if not harvested. Male and female flowers are borne in separate clusters on the same plant. Both types of flowers are small and inconspicuous, without petals; the calyx is light yellow or greenish yellow in color. A small fruit is formed from a fertilized female flower; this fruit was not utilized as food or for propagating purposes by the Hawaiians. Instead, small "seed tubers" ("seedling" or "baby" secondary yams that form in addition to the main tuber) were planted.

Uhiuhi
Caesalpinia kavaiensis H. Mann
(endemic)

Uhiuhi is a native Hawaiian leguminous tree up to thirty feet tall with rough, scaly bark and bipinnate (twice-divided) leaves. The pinkish purple to red flowers are borne in terminal clusters one to four inches long. The pods are broad and thin, winged, three to four inches long, and bear two to four seeds. The wood is very hard, dense, and nearly black. It is officially listed as an endangered species by the U.S. Fish and Wildlife Service and is thus protected by federal as well as State of Hawaii statutes. *Uhiuhi* is now found only on Kaua'i, O'ahu, and Maui, but early reports also place it on Hawai'i.

PLATE 92
'Uki
Machaerina angustifolia (Gaud.) T. Koyama
(indigenous)

'Uki is a perennial with stems ten to twenty inches high with leaves borne primarily at the base with a single leaf about midway up the stem. Leaves are narrow, pale green, and smooth. The inflorescence is a large panicle bearing four to five partial panicles, the panicles branched and subtended by a brown bract sheath. Flowers, borne on spikelets, are either perfect or staminate.

Plate 93
'Uki'uki
Dianella sandwicensis Hook. & Arnott
(indigenous)

'Uki'uki belongs to the lily family and has long, narrow, smooth, somewhat leathery leaves with close, longitudinally parallel veins. The leaves are one to three feet long by 1/2 to one inch wide. They form a two-rowed cluster along a short erect stem. From the latter arises a branching flower cluster; the flowers are small and bluish. The fruits are long-persistent berries, a conspicuous light or dark blue.

PLATE 94
ʻŪlei or *Eluehe*
Osteomeles anthyllidifolia (Sm.) Lindl.
(indigenous)

ʻŪlei, a native of Hawaiʻi, is a member of the rose family, and grows in dry parts of the six main islands at altitudes between sea level and 8,000 feet. Although it becomes widely spreading over the ground, or rises to one to two feet in protected areas, it also grows as large erect shrubs or even small trees. It bears compound leaves; these consist of eleven to twenty-five oblong leaflets 1/2 inch long that are arranged in pairs, with an extra leaflet at the tip. When young, the leaves appear silvery because they bear silky hairs, but in time the upper surface becomes smooth. The white roselike flowers, loosely clustered at the tips of branches, are 1/2 inch in diameter; they have five petals, five sepals, and many stamens. The fruits, about 1/3 inch in diameter, are globose and white, with a white to purple, sweet-tasting pulp; this contains five hard seeds called stones.

PLATE 95
'Ulu (breadfruit)
Artocarpus altilis (S. Parkinson ex Z) Fosb.
(Polynesian-introduced)

'Ulu is a most attractive spreading tree, with heights ranging from thirty to fifty feet and a trunk diameter up to 1-1/2 to two feet. The foliage is luxuriant and beautiful, consisting of very large leathery leaves, from one to three feet long; they are dark green and the upper side is glossy. The margins vary with the variety; the "Hawaiian variety" (the one brought to Hawai'i by the original Polynesian settlers) has margins that are slightly serrated and cut rather deeply into blunt lobes. Young leaves, "accordian-pleated," are protected inside a sheath until "ready" to expand. The male and female inflorescenses are borne separately on the same tree. The male inflorescence (*ule,* penis, which it resembles), made up of thousands of minute, staminate, closely crowded flowers with pollen, is a stiff, club-shaped "spike" measuring six to eight inches long and about an inch in diameter. Its color is yellowish or chartreuse when young, turning with age to beige, tan, and, finally, to dark brown when it falls. The female inflorescence forms a spear consisting of hundreds of closely appressed pistillate flowers that are attached to a "core," which is an extension of the stem of the flower cluster. The individual flowers develop into fruitlets that press against each other as they expand, changing from a rounded to an angular shape; together they form a composite or collective fruit, known botanically as a sorosis. The fruiting season is usually May to September. The fruit of the Hawaiian variety is slightly oblong, about six inches in diameter, and it weighs two to three pounds. It has a tough, wartylike rind, sometimes called skin, enclosing a somewhat mealy flesh surrounding a club-shaped "core," an extension of the fruit stem. The color of the rind of the immature fruit is a bright pea green, and the flesh at this stage of development is hard, white, and starchy. At maturity, the rind turns a tannish, moss green color, and a certain amount of milky sap oozes from around the fruit stem; these changes indicate that it is time to pick the fruit, even though it feels "as hard as a rock." On standing, the fruit ripens and becomes soft in twenty-four to forty-eight hours: the starch turns to sugar and the flesh is sweet. The Hawaiian variety bears no seeds.

PLATE 96

Uluhe (false staghorn fern)

Dicranopteris linearis (N. L. Burm.) Underw.

(endemic)

Uluhe has fronds that fork repeatedly an indefinite number of times, bearing a pair of pinnae, six to nine inches by two to four inches, at each fork as well as at the tip; these are cut into narrow lobes about 1/2 inch wide. Between each fork is a dormant bud. The pinnae may have a brownish wool layer on the underside that may be irregularly spotted with uncovered round fruit dots (sori).

Plate 97
Wauke (paper mulberry)
Broussonetia papyrifera (L.) Venten.
(Polynesian-introduced)

Wauke, ordinarily a shrub or a small tree, can grow into a tree as tall as fifty feet if allowed to grow indefinitely. When grown singly (isolated), it branches profusely. Shoots or suckers form readily on roots, and such root shoots may appear above the ground at considerable distances from the mother plant. The outer bark of the stem and branches is grayish green and smooth. The leaves are roughly heart-shaped, with either entire margins or they are two- or three-lobed. A plant may have all forms of leaves, two or three, or only one. The upper surface of the leaves is rough, like sandpaper; the underside is woolly or downy. They are from six to eight inches long, and from four to six inches wide at their greatest dimension. The leaf stem is usually short, about an inch long, but may be as long as six inches. Large stipules are present at the base of these stems. Male and female flowers are borne on separate plants. The female flowers form a round head, about an inch in diameter, fuzzy because the stigmas are long; hairy bracts are present. The fruit that forms is orange when mature; it is edible but not palatable. A close relative, the white mulberry, *Morus alba* L., was grown in Hawaiʻi for its fruit, and the leaves were used as food for silkworms.

PLATE 98
Wiliwili
Erythrina sandwicensis Degener
(endemic)

Wiliwili is a native Hawaiian leguminous tree found primarily in dry areas, on dry coral plains, and on lava flows, from sea level to 2,000 feet. Trees are eighteen to thirty feet high, wide spreading; the trunk is sometimes short and thick. The branches are somewhat spiny, gnarled, and yellowish hairy at their tips. The leaves are long-stemmed, with three ovate, leathery leaflets, hairy on the lower surface, each about two to 2-1/2 inches long and three to 3-1/2 inches wide. The leaves of some fall late in the season, with new ones appearing in spring when or after the flowers open. The wood is very light when dry and is often compared to balsa wood. Flowers are borne in clusters near the tip of branches. They have the general form of leguminous flowers and range in color from pale red through oranges and deep reds to white, with yellow, chartreuse, and pale green as variations. The pod ordinarily contains one to three red or orange oblong seeds.

SOURCES FOR DESCRIPTION OF PLANTS

Degener (and Degener) (1932–1980); Handy (1940); Handy and Handy (1972); Hillebrand (1888); Kimura and Nagata (1980); Lamoureux (1976); Neal (1948); Rock (1913); Wagner et al. (1990).

GLOSSARY

'a'ali'i, Dodonaea viscosa Jacq.
'a'ali'i mahu, pūkiawe, kānehoa, kāwa'u, maiele, Styphelia tameiameiae (Cham. & Schlechtend.) F. v. Muell.
Acacia koa A. Gray, *koa*
Acacia koaia Hillebr., *koai'a, koai'e*
a'e, mānele, soapberry, *Sapindus saponaria* L.
Agrostis avenacea J. G. Gmelin, *he'upueo*
'ahakea, Bobea spp.
'āheahea, 'āweoweo, native pigweed, lamb's quarters, *Chenopodium oahuense* (Meyen) Aellen
'ahu'awa, Mariscus javanicus (Houtt.) Merr. & Metcalf
'aiakanēnē, kūkaenēnē, Coprosma ernodeoides A. Gray
'aiea, Nothocestrum spp.
'aka'akai, great bulrush, *Schoenoplectus lacustris* (L.) Palla
'ākala, Hawaiian raspberry, *Rubus hawaiensis* A. Gray, *R. macraei* A. Gray
'ākia, Wikstroemia spp.
'akoko, Chamaesyce spp.
'akū'akū, Cyanea platyphylla (A. Gray) Hillebr.
'āla'a, Pouteria sandwicensis (A. Gray) Baehni & Degener
'ala'ala wai nui, peperomia, *Peperomia* spp.
alahe'e, Psydrax odorata (G. Forster) A. C. Smith & S. Darwin
Aleurites moluccana (L.) Willd., *kukui,* candlenut tree
ali'ipoe, wild canna, *Canna indica* L.
aloalo, native hibiscus, *Hibiscus* spp.
Alocasia macrorrhiza (L.) Schott, *'ape*
Alphitonia ponderosa Hillebr., *kauila, kauwila, o'a*
Alyxia oliviformis Gaud., *maile*
'ama'u, Sadleria cyatheoides Kaulf.
Antidesma spp., *hame, ha'ā, mehame*
'ape, Alocasia macrorrhiza (L.) Schott
Argemone glauca (Nutt. ex Prain) Pope, *pua kala,* prickly poppy
arrowroot, Polynesian; *pia; Tacca leontopetaloides* (L.) Kuntze
Artocarpus altilis (S. Parkinson ex Z) Fosb., *'ulu,* breadfruit
Asparagopsis taxiformis (Delile) Trevisan, *limu kohu*
Astelia spp., *pa'iniu*
Athyrium arnotii (ined.), *hō'i'o*
āulu, kaulu, Pouteria sandwicensis (A. Gray) Baehni & Degener
'awa, Piper methysticum G. Forster
'awapuhi, 'awapuhi kuahiwi, shampoo ginger, *Zingiber zerumbet* (L.) Sm.
'awapuhi kuahiwi, 'awapuhi, shampoo ginger, *Zingiber zerumbet* (L.) Sm.
'āweoweo, 'āheahea, native pigweed, lamb's quarters, *Chenopodium oahuense* (Meyen) Aellen
'āwikiwiki, puakauhi, Canavalia spp.
bamboo, *'ohe, Schizostachyium glaucifolium* (Rupr.) Munro
banana, *mai'a, Musa* spp.
beardgrass, twisted; tanglehead; *pili; Heteropogon contortus* (L.) P. Beauv. ex Roem. & Schult
Bidens spp., *ko'oko'olau*
Bobea spp., *'ahakea*
breadfruit, *'ulu, Artocarpus altilis* (S. Parkinson ex Z) Fosb.
Broussonetia papyrifera (L.) Venten., *wauke,* paper mulberry
bulrush, great; *'aka'akai; Schoenoplectus lacustris* (L.) Palla
Caesalpinia kavaiensis H. Mann, *uhiuhi*
calabash tree, *la'amia, Crescentia cujete* L.
Calophyllum inophyllum L., *kamani*
Canavalia spp., *'āwikiwiki, puakauhi*
candlenut tree, *kukui, Aleurites moluccana* (L.) Willd.
canna, wild; *ali'ipoe; Canna indica* L.
Canna indica L., *ali'ipoe,* wild canna
Capparis sandwichiana DC, *maiapilo, pilo, pua pilo*
Chamaesyce spp., *'akoko*
Charpentiera spp., *pāpala*
chaste tree, beach vitex, *pōhinahina, kolokolo kahakai, Vitex rotundifolia* L. fil.
Cheirodendron spp., *'ōlapa, lapalapa*
Chenopodium oahuense (Meyen) Aellen, *'āheahea, 'āweoweo,* native pigweed, lamb's quarters
Cibotium glaucum (Sm.) Hook. & Arnott, *hāpu'u pulu,* tree fern
Cibotium menziesii Hook., *hāpu'u 'i'i,* tree fern
Cladium jamaicense Crantz, *'uki*
Clermontia spp., *'ōhā, 'ōhā wai, hāhā*

coconut, *niu, Cocos nucifera* L.
Cocos nucifera L., *niu,* coconut
Colocasia esculenta (L.) Schott, *kalo,* taro
Colubrina oppositifolia Brongn. ex H. Mann, *kauila, kauwila*
Coprosma ernodeoides A. Gray, *'aiakanēnē, kūkaenēnē*
Cordia subcordata Lam., *kou*
Cordyline fruticosa (L.) A. Chev., *kī, ti*
cotton, Hawaiian; *ma'o; Gossypium tomentosum* Nutt. ex Seem.
Crescentia cujete L., *la'amia,* calabash tree
Curcuma longa L., *'ōlena,* turmeric
Cuscuta sandwichiana Choisy, *kauna'oa*
Cyanea platyphylla (A. Gray) Hillebr., *'akū'akū*
Cyperus laevigatus L., *makaloa*
Dianella sandwicensis Hook. & Arnott, *'uki'uki*
Dicranopteris linearis (N. L. Burm.) Underw., *uluhe,* false staghorn fern
Digitaria setigera Roth, *kūkaepua'a*
Dioscorea alata L., *uhi,* yam
Dioscorea bulbifera L., *hoi,* bitter yam
Dioscorea pentaphylla L., *pi'a,* wild yam
Diospyros spp., *lama*
Dodonaea viscosa Jacq., *'a'ali'i*
Dryopteris spp., *kikawaiō, pakikawaiō*
Dubautia spp., *kūpaoa, na'ena'e*
eluehe, 'ūlei, Osteomeles anthyllidifolia (Sm.) Lindl.
Eragrostis variabilis (Gaud.) Steud., *kāwelu*
Erythrina sandwicensis Degener, *wiliwili*
Eugenia spp., *nīoi*
fern, false staghorn; *uluhe; Dicranopteris linearis* (N. L. Burm.) Underw.
fern, lace; *pala'ā; Sphenomeris chinensis* (L.) Maxon
fern, *maile*-scented; *laua'e; Phymatosorus scolopendria* (N. L. Burm.) Pic.-Ser.
Freycinetia arborea Gaud., *'ie'ie,* climbing pandanus
gardenia, native; *nānū, nā'ū; Gardenia* spp.
Gardenia spp., *nānū, nā'ū,* native gardenia; *Gardenia brighamii* H. Mann; *Gardenia mannii* St. John & Kuykendall; *Gardenia remyi* H. Mann
ginger, shampoo; *'awapuhi, 'awapuhi kuahiwi; Zingiber zerumbet* (L.) Sm.
Gossypium tomentosum Nutt. ex Seem., *ma'o,* Hawaiian cotton
gourd plant, *pōhue, ipu, Lagenaria siceraria* (Molina) Standl.
hā, 'ōhi'a hā, Syzygium sandwicensis (A. Gray) Nied.
ha'ā, hame, mehame, Antidesma spp.
hāhā, 'ōhā, 'ōhā wai, Clermontia spp.
hala, pū hala, screw pine, *Pandanus tectorius* S. Parkinson ex Z
hala pepe, Pleomele spp.
hame, ha'ā, mehame, Antidesma spp.
hāpu'u 'i'i, tree fern, *Cibotium menziesii* Hook.
hāpu'u pulu, tree fern, *Cibotium glaucum* (Sm.) Hook. & Arnott
hau hele 'ula, koki'o, Kokia spp.
hau, Hibiscus tiliaceus L.
hea'e, kāwa'u, rue, *Zanthoxylum* spp.
Heliotropium anomalum Hook. & Arnott var. *argenteum* A. Gray, *hinahina, hinahina kū kahakai*
Heteropogon contortus (L.) P. Beauv. ex Roem. & Schult, *pili,* twisted beardgrass, tanglehead
he'upueo, Agrostis avenacea J. G. Gmelin
hibiscus, native; *aloalo; Hibiscus* spp.
Hibiscus spp., *aloalo,* native hibiscus
Hibiscus tiliaceus L., *hau*
hinahina, hinahina kū kahakai, Heliotropium anomalum Hook. & Arnott var. *argenteum* A. Gray
hinihina kū kahakai, hinahina, Heliotropium anomalum Hook. & Arnott var. *argenteum* A. Gray
hō'awa, Pittosporum spp.
hoi, bitter yam, *Dioscorea bulbifera* L.
hō'i'o, Athyrium arnotii (ined.)
hōlei, Ochrosia spp.
holly, *kāwa'u, Ilex anomala* Hook. & Arnott
hōpue, ōpuhe, Urera glabra (Hook. & Arnott) Wedd.
hua lewa, unidentified plant
'ie'ie, climbing pandanus, *Freycinetia arborea* Gaud.
'ihi, Portulaca spp.
Ilex anomala Hook. & Arnott, *kāwa'u,* holly
'iliahi, sandalwood, *Santalum* spp.
'ilima, Sida fallax Walp.
Ipomoea batatas (L.) Lam., *'uala,* sweet potato
Ipomoea indica (J. Burm.) Merr., *koali 'awa, koali 'awahia,* morning glory
Ipomoea pes-caprae (L.) R. Br., subsp. *brasiliensis* (L.) Ooststr., morning glory
Ipomoea spp., *'uala koali,* sweet-potato morning glory
ipu, pōhue, gourd plant, *Lagenaria siceraria* (Molina) Standl.
kā 'i'iwi, nuku 'i'iwi, Strongylodon ruber Vogel
kalamona, kolomona, Senna gaudichaudii (Hook. & Arnott) H. Irwin & Barneby
kalo, taro, *Colocasia esculenta* (L.) Schott
kāmakahala, Labordia spp.
kamani, Calophyllum inophyllum L.
kānehoa, pūkiawe, 'a'ali'i mahu, kāwa'u, maiele, Styphelia tameiameiae (Cham. & Schlechtend.) F. v. Muell.
kauila, o'a, kauwila, Alphitonia ponderosa Hillebr.
kaulu, Pouteria sandwicensis spp. (A. Gray) Baehni & Degener (sapodilla family)
kauna'oa, Cuscuta sandwichiana Choisy
kauwila, kauila, Alphitonia ponderosa Hillebr., *Colubrina oppositifolia* Brongn. ex H. Mann

kāwa'u, hea'e, rue, *Zanthoxylum* spp.
kāwa'u, holly, *Ilex anomala* Hook. & Arnott
kāwa'u, pūkiawe, 'a'ali'i mahu, kānehoa, maiele, Styphelia tameiameiae (Cham. & Schlechtend.) F. v. Muell.
kāwelu, Eragrostis variabilis (Gaud.) Steud.
kī, ti, *Cordyline fruticosa* (L.) A. Chev.
kikawaiō, pakikawaiō, Dryopteris spp.
kō, sugarcane, *Saccharum officinarum* L.
koa, Acacia koa A. Gray
koai'a, koai'e, Acacia koaia Hillebr.
koai'e, koai'a, Acacia koaia Hillebr.
koali, wild morning glory, *Ipomoea* spp.
koali 'awa, koali 'awahia, morning glory, *Ipomoea indica* (J. Burm.) Merr.
koali 'awahia, koali 'awa, morning glory, *Ipomoea indica* (J. Burm.) Merr.
Kokia spp., *koki'o, hau hele 'ula*
koki'o, hau hele 'ula, Kokia spp.
kōlea, Myrsine spp.
kolekole lehua, kolokolo kuahiwi, Lysimachia spp.
kolokolo kahakai, pōhinahina, beach vitex, chaste tree, *Vitex rotundifolia* L. fil.
kolokolo kuahiwi, kolekole lehua, Lysimachia spp.
kolomona, kalamona, Senna gaudichaudii (Hook. & Arnott) H. Irwin & Barneby
ko'oko'olau, Bidens spp.
kou, Cordia subcordata Lam.
kualohia, unidentified grass or sedge
kūkaenēnē, 'aiakanēnē, Coprosma ernodeoides A. Gray
kūkaepua'a, Digitaria setigera Roth
kukui, candlenut tree, *Aleurites moluccana* (L.) Willd.
kūpaoa, na'ena'e, Dubautia spp.
la'amia, calabash tree, *Crescentia cujete* L.
Labordia spp., *kāmakahala*
Lagenaria siceraria (Molina) Standl., *pōhue, ipu,* gourd plant
lama, Diospyros spp.
lamb's quarters, *'āheahea, 'āweoweo, Chenopodium oahuense* (Meyen) Aellen
lapalapa, 'ōlapa, Cheirodendron spp.
laua'e, maile-scented fern, *Phymatosorus scolopendria* (N. L. Burm.) Pic.-Ser.
laukahi kuahiwi, plantain, *Plantago* spp.
lehua, 'ōhi'a lehua, Metrosideros polymorpha Gaud., *M. macropus* Hook. & Arnott
lehua 'āhihi, Metrosideros tremuloides (A. Heller) P. Knuth
lehua papa, Metrosideros rugosa A. Gray
limu kala, Sargassum echinocarpum, J. G. Agardh
limu kohu, Asparagopsis taxiformis (Delile) Trevisan
Lipochaeta spp., *nehe*
lonomea, Sapindus oahuensis Hillebr. ex Radlk.
loulu, fan palm, *Pritchardia* spp.
Lycopodium cernuum L., *wāwae'iole*
Lysimachia spp., *kolokolo kuahiwi, kolekole lehua*
ma'aloa, 'oloa, ma'oloa, Neraudia melastomifolia Gaud.
mā'auea, unidentified plant
Machaerina angustifolia (Gaud.) T. Koyama, *'uki*
mai'a, banana, *Musa* spp.
maiapilo, pilo, pua pilo, Capparis sandwichiana DC
maiele, pūkiawe, 'a'ali'i mahu, kānehoa, kāwa'u, Styphelia tameiameiae (Cham. & Schlechtend.) F. v. Muell.
maile, Alyxia oliviformis Gaud.
makaloa, Cyperus laevigatus L.
Malay apple, *'ōhi'a 'ai, Syzygium malaccense* (L.) Merr. & Perry
māmake, māmaki, Pipturus spp.
māmaki, māmake, Pipturus spp.
māmane, mamani, Sophora chrysophylla (Salisb.) Seem.
mamani, māmane, Sophora chrysophylla (Salisb.) Seem.
mānele, a'e, soapberry, *Sapindus saponaria* L.
ma'o, Hawaiian cotton, *Gossypium tomentosum* Nutt. ex Seem.
ma'oloa, ma'aloa, 'oloa, Neraudia melastomifolia Gaud.
Marattia douglasii (Presl) Baker, *pala*
Mariscus javanicus (Houtt.) Merr. & Metcalf, *'ahu'awa*
mehame, hame, ha'ā, Antidesma spp.
Metrosideros macropus Hook. & Arnott, *'ōhi'a lehua, lehua*
Metrosideros polymorpha Gaud., *'ōhi'a lehua, lehua*
Metrosideros rugosa A. Gray, *lehua papa*
Metrosideros tremuloides (A. Heller) P. Knuth, *lehua 'āhihi*
Microlepia strigosa (Thunb.) K. Presl, *palai, palapalai*
milo, portia tree, *Thespesia populnea* (L.) Sol. ex Corrêa
moa, Psilotum nudum (L.) P. Beauv.
mokihana, Pelea anisata H. Mann
Morinda citrifolia L., *noni,* Indian mulberry
morning glory, *koali 'awa, koali 'awahia, Ipomoea indica* (J. Burm.) Merr.
morning glory, beach; *pōhuehue; Ipomoea pes-caprae* (L.) R. Br.
morning glory, sweet potato; *'uala koali, Ipomoea* spp.
morning glory, wild; *koali; Ipomoea* spp.
mountain apple, *'ōhi'a 'ai, Syzygium malaccense* (L.) Merr. & Perry
mulberry, Indian; *noni; Morinda citrifolia* L.
mulberry, paper; *wauke Broussonetia papyrifera* (L.) Venten.
Musa spp., *mai'a,* banana
Myoporum sandwicense A. Gray, *naio,* bastard sandalwood
Myrsine spp., *kōlea*
na'ena'e, kūpaoa, Dubautia spp.
naio, bastard sandalwood, *Myoporum sandwicense* A. Gray
nānū, nā'ū, native gardenia, *Gardenia* spp.

Nasturtium sarmentosum (G. Forster ex DC) Schinz & Guillaumin, *pā'ihi*
nā'ū, nānū, native gardenia, *Gardenia* spp.
naupaka, beach; *naupaka kahakai; Scaevola sericea* Vahl
naupaka, mountain; *naupaka kuahiwi; Scaevola* spp.
naupaka kahakai, beach *naupaka, Scaevola sericea* Vahl
naupaka kuahiwi, mountain *naupaka, Scaevola* spp.
nehe, Lipochaeta spp.
Neraudia melastomifolia Gaud., *ma'aloa, 'oloa, ma'oloa*
Nestegis sandwicensis (A. Gray) Degener, I. Degener & L. Johnson, *olopua, pua*
nightshade, glossy; *pōpolo; Solanum americanum* Mill.
nīoi, Eugenia spp.
niu, coconut, *Cocos nucifera* L.
noni, Indian mulberry, *Morinda citrifolia* L.
Nothocestrum spp., *'aiea*
nuku 'i'iwi, kā 'i'iwi, Strongylodon ruber Vogel
o'a, kauila, kauwila, Alphitonia ponderosa Hillebr.
Ochrosia spp., *hōlei*
'ōhā, 'ōhā wai, hāhā, Clermontia spp.
'ōhā wai, 'ōhā, hāhā, Clermontia spp.
'ohai, Sesbania tomentosa Hook. & Arnott
'ohe, bamboo *Schizostachyium glaucifolium* (Rupr.) Munro
'ohe makai, Reynoldsia sandwicensis A. Gray
'ōhelo, Vaccinium spp.
'ōhi'a 'ai, mountain apple, Malay apple, *Syzygium malaccense* (L.) Merr. & Perry
'ōhi'a hā, hā, Syzygium sandwicensis (A. Gray) Nied.
'ōhi'a lehua, lehua, Metrosideros polymorpha Gaud., *M. macropus* Hook. & Arnott
'ōlapa, lapalapa, Cheirodendron spp.
'ōlena, turmeric, *Curcuma longa* L.
'oloa, ma'aloa, ma'oloa, Neraudia melastomifolia Gaud.
olonā, Touchardia latifolia Gaud.
olopua, pua, Nestegis sandwicensis (A. Gray) Degener, I. Degener & L. Johnson
ōpuhe, hōpue, Urera glabra (Hook. & Arnott) Wedd.
Osteomeles anthyllidifolia (Sm.) Lindl., *'ūlei, eluehe*
pā'ihi, Nasturtium sarmentosum (G. Forster ex DC) Schinz & Guillaumin
pa'iniu, Astelia spp.
pakikawaiō, kikawaiō, Dryopteris spp.
pala, Marattia douglasii (Presl) Baker
pala'ā, lace fern, *Sphenomeris chinensis* (L.) Maxon
palai, palapalai, Microlepia strigosa (Thunb.) K. Presl
palapalai, palai, Microlepia strigosa (Thunb.) K. Presl
palm, fan; *loulu; Pritchardia* spp.
pandanus, climbing; *'ie'ie; Freycinetia arborea* Gaud.
Pandanus tectorius S. Parkinson ex Z, *hala, pū hala,* screw pine
pāpala, Charpentiera spp.
Pelea anisata H. Mann, *mokihana*
peperomia, *'ala'ala wai nui, Peperomia* spp.
Peperomia spp., *'ala'ala wai nui,* peperomia
Phymatosorus scolopendria (N. L. Burm.) Pic.-Ser., *laua'e, maile*-scented fern
Phytolacca sandwicensis Endl., *pōpolo kū mai,* native pokeberry
pia, Polynesian arrowroot, *Tacca leontopetaloides* (L.) Kuntze
pi'a, wild yam, *Dioscorea pentaphylla* L.
pigweed, native; *'āheahea, 'āweoweo; Chenopodium oahuense* (Meyen) Aellen
pili, twisted beardgrass, tanglehead, *Heteropogon contortus* (L.) P. Beauv. ex Roem. & Schult
pilo, maiapilo, pua pilo, Capparis sandwichiana DC
Piper methysticum G. Forster, *'awa*
Pipturus spp., *māmaki, māmake*
Pittosporum spp., *hō'awa*
Plantago spp., *laukahi kuahiwi,* plantain
plantain, *laukahi kuahiwi, Plantago* spp.
Pleomele spp., *hala pepe*
pōhinahina, kolokolo kahakai, beach vitex, chaste tree, *Vitex rotundifolia* L. fil.
pōhue, ipu, gourd plant, *Lagenaria siceraria* (Molina) Standl.
pōhuehue, beach morning glory, *Ipomoea pes-caprae* (L.) R. Br., subsp. *brasiliensis* (L.) Ooststr.
pokeberry, native; *pōpolo kū mai; Phytolacca sandwicensis* Endl.
pōpolo, glossy nightshade, *Solanum americanum* Mill.
pōpolo kū mai, native pokeberry, *Phytolacca sandwicensis* Endl.
poppy, prickly; *pua kala; Argemone glauca* (Nutt. ex Prain) Pope
portia tree, *milo, Thespesia populnea* (L.) Sol. ex Corrêa
Portulaca spp., *'ihi*
Pouteria sandwicensis (A. Gray) Baehni & Degener, *'āla'a, āulu, kaulu*
Pritchardia spp., *loulu,* fan palm
Psilotum nudum (L.) P. Beauv., *moa*
Psydrax odorata (G. Forster) A. C. Smith & S. Darwin
pū hala, hala, screw pine, *Pandanus tectorius* S. Parkinson ex Z
pua, olopua, Nestegis sandwicensis (A. Gray) Degener, I. Degener & L. Johnson
pua kala, prickly poppy, *Argemone glauca* (Nutt. ex Prain) Pope
pua pilo, pilo, maiapilo, Capparis sandwichiana DC
puakauhi, 'āwikiwiki, Canavalia spp.
pūkiawe, 'a'ali'i mahu, kānehoa, kāwa'u, maiele, Styphelia tameiameiae (Cham. & Schlechtend.) F. v. Muell.
raspberry, Hawaiian; *'ākala; Rubus hawaiensis* A. Gray, *R. macraei* A. Gray
Reynoldsia sandwicensis A. Gray, *'ohe makai*
Rubus hawaiensis A. Gray, *'ākala,* Hawaiian raspberry

Rubus macraei A. Gray, *'ākala,* Hawaiian raspberry
rue, *kāwa'u, hea'e, Zanthoxylum* spp.
Saccharum officinarum L., *kō,* sugarcane
Sadleria cyatheoides Kaulf., *'ama'u*
sandalwood, *'iliahi, Santalum* spp.
sandalwood, bastard; *naio; Myoporum sandwicense* A. Gray
Santalum spp., *'iliahi,* sandalwood
Sapindus oahuensis Hillebr. ex Radlk., *lonomea*
Sapindus saponaria L., *a'e, mānele,* soapberry
Sargassum echinocarpum, limu kala, J. G. Agardh
Scaevola sericea Vahl, *naupaka kahakai,* beach *naupaka*
Scaevola spp., *naupaka kuahiwi,* mountain *naupaka*
Schizostachyium glaucifolium (Rupr.) Munro, *'ohe,* bamboo
Schoenoplectus lacustris (L.) Palla, *'aka'akai,* great bulrush
screw pine, *hala, pū hala, Pandanus tectorius* S. Parkinson ex Z
Senna gaudichaudii (Hook. & Arnott) H. Irwin & Barneby, *Kolomona, kalamona*
Sesbania tomentosa Hook. & Arnott, *'ohai*
Sida fallax Walp., *'ilima*
soapberry, *a'e, mānele, Sapindus saponaria* L.
Solanum americanum Mill., *pōpolo,* glossy nightshade
Sophora chrysophylla (Salisb.) Seem., *māmane, manani*
Sphenomeris chinensis (L.) Maxon, *pala'ā,* lace fern
Strongylodon ruber Vogel, *nuku 'i'iwi, kā 'i'iwi*
Styphelia tameiameiae (Cham. & Schlechtend.) F. v. Muell., *pūkiawe, 'a'ali'i mahu, kānehoa, kāwa'u, maiele*
sugarcane, *kō, Saccharum officinarum* L.
sweet potato, *'uala, Ipomoea batatas* (L.) Lam.
Syzygium malaccense (L.) Merr. & Perry, *'ōhi'a 'ai,* mountain apple, Malay apple
Syzygium sandwicensis (A. Gray) Nied., *'ōhi'a hā, hā*
Tacca leontopetaloides (L.) Kuntze, *pia,* Polynesian arrowroot
tanglehead, twisted beardgrass, *pili, Heteropogon contortus* (L.) P. Beauv. ex Roem. & Schult
taro, *kalo, Colocasia esculenta* (L.) Schott
Thespesia populnea (L.) Sol. ex Corrêa, *milo,* portia tree
ti, kī, Cordyline fruticosa (L.) A. Chev.
Touchardia latifolia Gaud., *olonā*
tree fern, *hāpu'u 'i'i, Cibotium menziesii* Hook.
tree fern, *hāpu'u pulu, Cibotium glaucum* (Sm.) Hook. & Arnott
turmeric, *'ōlena, Curcuma longa* L.
'uala, sweet potato, *Ipomoea batatas* (L.) Lam.
'uala koali, sweet-potato morning glory, *Ipomoea* spp.
'uhaloa, Waltheria indica L.
uhi, yam, *Dioscorea alata* L.
uhiuhi, Caesalpinia kavaiensis H. Mann
'uki, Cladium jamaicense Crantz, *Machaerina angustifolia* (Gaud.) T. Koyama
'uki'uki, Dianella sandwicensis Hook. & Arnott
'ūlei, eluehe, Osteomeles anthyllidifolia (Sm.) Lindl.
'ulu, breadfruit, *Artocarpus altilis* (S. Parkinson ex Z) Fosb.
uluhe, false staghorn fern, *Dicranopteris linearis* (N. L. Burm.) Underw.
Urera glabra (Hook. & Arnott) Wedd., *hōpue, ōpuhe*
Vaccinium spp., *ōhelo*
vitex, beach; chaste tree; *pōhinahina, kolokolo kahakai; Vitex rotundifolia* L. fil.
Vitex rotundifolia L. fil., *pōhinahina, kolokolo kahakai,* beach vitex, chaste tree
Waltheria indica L., *'uhaloa*
wauke, paper mulberry, *Broussonetia papyrifera* (L.) Venten.
wāwae'iole, Lycopodium cernuum L.
Wikstroemia spp., *'ākia*
wiliwili, Erythrina sandwicensis Degener
yam, *uhi, Dioscorea alata* L.
yam, bitter; *hoi; Dioscorea bulbifera* L.
yam, wild; *pi'a; Dioscorea pentaphylla* L.
Zanthoxylum spp., *kāwa'u, hea'e,* rue
Zingiber zerumbet (L.) Sm., *'awapuhi, 'awapuhi kuahiwi,* shampoo ginger

BIBLIOGRAPHY

Abbott, Isabella A. 1982. The ethnobotany of Hawaiian taro. *Native Planters (Ho'okupu kalo)* 1 (1): 17–22.

Abbott, Isabella A., and Eleanor H. Williamson. 1974. *Limu. An ethnobotanical study of some edible Hawaiian seaweeds.* Lawai, Kauai, Hawaii: Pacific Tropical Botanical Garden.

Alexander, Arthur C. 1891. "A brief history of land titles in the Hawaiian kingdom." *Hawaiian Annual.* Honolulu.

Anonymous. 1939. *An historic inventory of the physical, social and economic and industrial resources of the Territory of Hawaii.* Honolulu.

Apple, Russel A. 1971. *The Hawaiian thatched house.* U.S. Department of Interior, National Park Service. Office of History and Historic Architecture. Western Service Center. San Francisco.

Barrère, Dorothy B., Mary Kawena Pukui, and Marion Kelley. 1980. *Hula.* Pacific Anthropological Records 30. Honolulu: Bernice Pauahi Bishop Museum Department of Anthropology.

Beamer, Nona. 1987. *Nā mele hula: a collection of Hawaiian hula chants.* Laie, Hawaii: The Institute for Polynesian Studies, Brigham Young University.

Beckley, Emma Metcalf. 1883. *Hawaiian fisheries and methods of fishing.* Honolulu: Advertiser Steam Print.

Beckwith, Martha W. 1970. *Hawaiian mythology.* Honolulu: University of Hawaii Press. Reprint.

Bennet, Wendell C. 1930. Hawaiian heiaus. Ph.D. diss., University of Hawaii, Honolulu.

———. 1931. *Archaeology of Kauai.* Honolulu: Bishop Museum Press.

Bishop, Marcia Brown. 1940. *Hawaiian life of the pre-European period.* Peabody Museum, Salem. Portland, Maine: The Southworth-Anthoensen Press.

Blake, Thomas. 1961. *Hawaiian surfing: The ancient and royal sport.* Flagstaff, Arizona: Northland Press.

Brigham, William T. 1906. Mat and basket weaving of the ancient Hawaiians. *Bernice Pauahi Bishop Museum Memoirs* 2 (1): 1–105.

———. 1908. The ancient Hawaiian house. *Bernice Pauahi Bishop Museum Memoirs* 2 (3): 1–194; plates I–XL.

———. 1911. Ka hana kapa: The making of bark-cloth in Hawaii. *Bernice Pauahi Bishop Museum Memoirs* Vol. 3. (Kraus reprint, 1976.)

Bryan, Edwin H., Jr. 1938. *Ancient Hawaiian life.* Honolulu: Advertising Publishing Co.

———. 1965. "Fiberwork." *In Ancient Hawaiian civilization,* E. S .C. Handy and others, 129–133. Rutland, Vermont: Charles E. Tuttle Co.

Buck, Peter H. 1957. *Arts and crafts of Hawaii.* Bernice P. Bishop Museum Special Publication 45. Honolulu. (Reprinted as separate title by the Museum, 1964.)

———. 1965. "Polynesian migration." In *Ancient Hawaiian civilization,* E. S. C. Handy and others, 23–34. Rutland, Vermont: Charles E. Tuttle Co.

Charlot, John. 1982. Wet taro in Hawaiian literature: two examples. *Native Planters (Ho'okupu kalo)* 1 (1): 31–34.

Chung, H. L., 1925. The sweet potato in Hawaii. *Hawaii Agricultural Experiment Station Bulletin* 50.

Cook, James. 1967. *The voyage of the Resolution and Discovery 1776–1780* (The journals of Captain James Cook in his voyage of discovery). Part 1. J. C. Beaglehole, ed. Cambridge: Cambridge University Press.

Degener, Otto (and Isa Degener). 1932–1980. *Flora Hawaiiensis. The new illustrated flora of the Hawaiian Islands.* Honolulu: Published privately.

Dickey, Lyle A. 1928. *String figures from Hawaii, including some from New Hebrides and Gilbert Islands.* Bernice P. Bishop Museum Bulletin 54.

Ellis, William. 1969. *Polynesian researches: Hawaii,* new ed. Rutland, Vermont: Charles E. Tuttle Co.

Emerson, Nathaniel. 1965. *Unwritten literature of Hawaii: The sacred songs of the hula, collected and translated with notes and an account of the hula.* Reprint. Rutland, Vermont: Charles E. Tuttle Co.

Emory, Kenneth P. 1965a. "Wooden utensils and implements." In *Ancient Hawaiian civilization,* E. S. C. Handy and others, 123–128. Rutland, Vermont: Charles E. Tuttle Co.

———. 1965b. "Sports, games, and amusements." In *Ancient Hawaiian civilization,* E. S. C. Handy and others, 145–157. Rutland, Vermont: Charles E. Tuttle Co.

———. 1965c. "Warfare." In *Ancient Hawaiian civilization,* E. S. C. Handy and others, 233–240. Rutland, Vermont: Charles E. Tuttle Co.

———. 1965d. "Navigation." In *Ancient Hawaiian civilization,* E. S. C. Handy and others, 241–249. Rutland, Vermont: Charles E. Tuttle Co.

———. 1965e. "Comments on Chapter 2, Polynesian migrations by Buck." In *Ancient Hawaiian civilizations,* E. S. C. Handy and others, 319–320. Rutland, Vermont: Charles E. Tuttle Co.

Finney, Ben R. 1977. Voyaging canoes and the settlement of Polynesia. *Science* (June 17): 1277–1285.

Finney, Ben and James D. Housten. 1966. *Surfing, the sport of Hawaiian Kings.* Rutland, Vermont: Charles E. Tuttle Co.

Fornander, Abraham. 1919. "Traditional stories." In *Hawaiian antiquities and folk-lore.* Bernice Pauahi Bishop Museum Memoirs 5 (3): 570–659, 668–687. Honolulu. (Kraus reprint, 1985.)

———. 1919–1920. "An account of cultivation"; "An account of fishing"; and "Relating to amusements." In *Hawaiian antiquities and folk-lore.* Bernice Pauahi Bishop Museum Memoirs 6:160–171; 172–191; and 192–217, respectively.

Gutmanis, June. 1977. *Kahuna la'au lapa'au.* Norfolk Island: Island Heritage.

Handy, E. S. Craighill. 1940. *The Hawaiian planter.* Vol. I. *His plants, methods and areas of cultivation.* Bernice P. Bishop Museum Bulletin 161. Honolulu. (Kraus reprint, 1971.)

Handy, E. S. Craighill, and Elizabeth Green Handy. 1972. *Native planters in old Hawaii: Their life, lore, and environment.* Bernice P. Bishop Museum Bulletin 233. Honolulu.

Handy, E. S. Craighill, and Mary Kawena Pukui. 1972. *The Polynesian family system in Ka'u, Hawaii,* new ed. Rutland, Vermont: Charles E. Tuttle Co.

Handy, E. S. Craighill, Mary Kawena Pukui, and Katherine Livermore. 1934. *Outline of Hawaiian physical therapeutics.* Bernice P. Bishop Museum Bulletin 126. Honolulu.

Handy, E. S. Craighill, and others. 1965. *Ancient Hawaiian civilization,* rev. ed. Rutland, Vermont: Charles E. Tuttle Co. A series of lectures delivered at the Kamehameha Schools by eighteen persons. (Originally published in 1933 by the Kamehameha Schools, Honolulu.)

Hillebrand, W. F. 1888. *Flora of the Hawaiian Islands.* London: Williams and Norgate; New York: B. Westermann & Co.

Hobbs, Jean. 1935. *Hawaii, a pageant of the soil.* Palo Alto, California: Stanford University Press.

Holmes, Tommy. 1981. *The Hawaiian canoe.* Hanalei, Kauai, Hawaii: Editions Limited.

Ihara, Violet K. 1979. *Hawaiian barkcloth or tapa.* Research Bulletin 1. State of Hawaii, Department of Education. Honolulu district, Office of District Superintendent (mimeo).

Jordan, David S., and Barton W. Evermann, 1973. *The shore fish of Hawaii.* Reprint. Rutland, Vermont: Charles E. Tuttle Co.

Judd, Albert F. 1965. "Trees and plants." In *Ancient Hawaiian civilization,* E. S. C. Handy and others, 277–285. Rutland, Vermont: Charles E. Tuttle Co.

Ka'aiakamanu, D. M., and J. K. Akina. 1922. *Hawaiian herbs of medicinal value* (transl. by Akaiko Akana). (Hawaii) Territorial Board of Health. Facsimile reprint. Rutland, Vermont: Charles E. Tuttle Co.

Kaeppler, Adrienne L. 1970. Feather Cloaks, ship captains, and lords. *Occasional Papers of the Bernice Pauahi Bishop Museum* Vol. 24, no. 6. Honolulu.

———. 1975. *The fabrics of Hawaii (bark cloth).* Leigh-on-Sea, England: F. Lewis, Ltd.

———. 1980. *Kapa, Hawaiian bark cloth.* Hilo, Hawaii: Boon Books.

Kahananui, Dorothy M. 1960. *Music of ancient Hawaii. A brief survey.* Honolulu: Privately printed.

Kamakau, Samuel M. 1964. *Ka po'e kahiko: The people of old.* Bernice P. Bishop Museum Special Publication 51. Honolulu.

———. 1976. *The works of the people of old: Na hana a ka po'e kahiko.* Bernice P. Bishop Museum Special Publication 61. Honolulu.

Kane, Herb Kawainui. 1971(?). "Canoes." Typewritten ms. 29 pages.

Kelsey, Theodore. 1941. *Hawaiian kuehu-treatment.* Honolulu: Gloria B. Thom.

Kepelino. 1932. *Kepelino's traditions of Hawaii.* Honolulu: Bishop Museum. Kraus reprint 1978.

Kimura, Bert Y., and Kenneth M. Nagata. 1980. *Hawaii's vanishing flora.* Honolulu: Oriental Publishing Co.

Kirch, Patrick V. 1985. *Feathered gods and fishhooks.* Honolulu: University of Hawaii Press.

Knipe, Rita. 1982. Hāloa: the rootstalk-sprouted. *Native Planters (Ho'okupu kalo)* 1 (1): 23–30.

Kooijman, Simon. 1972. *Tapa in Polynesia.* Bernice P. Bishop Museum Bulletin 234. Honolulu.

Lamoureux, Charles H. 1976 *Trailside plants of Hawaii's National Parks.* Hawaii Natural History Association and National Park Service, U.S. Department of Interior.

Larsen, Nils P. 1965. "Ancient Hawaiian medical practices viewed by a doctor." In *Ancient Hawaiian civilization,* E. S. C. Handy and others, 260–267. Rutland, Vermont: Charles E. Tuttle Co.

Lawrence, Mary S. 1912. *Old time Hawaiians and their work.* Boston: Ginn and Co.

Lewis, D. 1972. *We the navigators.* Honolulu: University of Hawaii Press.

Luquiens, Huc-M. 1965. "Carving." In *Ancient Hawaiian civilization,* E. S. C. Handy and others, 225–232. Rutland, Vermont: Charles E. Tuttle Co.

McAllister, J. Gilbert. 1933a. *Archaeology of Oahu.* Bernice P. Bishop Museum Bulletin 104. Honolulu.

———. 1933b. *Archaeology of Kahoolawe.* Bernice P. Bishop Museum Bulletin 115. Honolulu.

McBride, L. R. 1972. *The kahuna.* Hilo, Hawaii: The Petroglyph Press.

———. 1975. *Practical folk medicine of Hawaii.* Hilo, Hawaii: The Petroglyph Press.

MacCaughey, Vaughan, and J. S. Emerson. 1913. The kalo in Hawaii. *Hawaiian Forester* 10:186–193, 225–231, 280–288, 315–323, 349–358, 371–375.

McDonald, Marie A. 1976. *Ka lei: The leis of Hawaii.* Honolulu: Topgallant Publishing Co.

Makekau, Abel. 1899. Kapa-beating. *Hawaii's Young People.* Vol. 3 (September 1898–June 1899).

Malo, David. 1951. *Hawaiian antiquities (Moolelo Hawaii),* 2d ed. Bernice P. Bishop Museum Special Publication 2. Honolulu.

Marques, A. J. B. 1914. Ancient Hawaiian music. *All about Hawaii* 40:97–107.

Marques, A. J. B., and Nathaniel B. Emerson. 1886. Music in Hawaii nei. *All about Hawaii* 11:51–60.

Maunupau, Thomas. 1965. "Aku and ahi fishing." In *Ancient Hawaiian civilization,* E. S. C. Handy and others, 105–111. Rutland, Vermont: Charles E. Tuttle Co.

Merlin, Mark. 1982. The origins and dispersals of true taro. *Native Planters (Ho'okupu kalo)* 1 (1): 6–16.

Miller, Carey D. 1957. "The influence of foods and food habits upon the stature and teeth of the ancient Hawaiians." In Charles E. Snow, *Early Hawaiians: An initial study of skeletal remains from Mokapu, Oahu.* 1974. Appendix E, 167–175. The University of Kentucky.

Mitchell, Donald K. 1972. *Resource units in Hawaiian culture,* rev. ed. Honolulu: Kamehameha Schools.

———. 1975. *Hawaiian games for today.* Honolulu: Kamehameha Schools Press.

Nagata, Kenneth. 1971. Hawaiian medicinal plants. *Economic Botany* 25 (3): 245–254.

Neal, Marie C. 1948. *In gardens of Hawaii.* Bernice P. Bishop Museum Special Publication 40. Honolulu.

Plews, Edith Rice. 1965. "Poetry." In *Ancient Hawaiian civilization,* E. S. C. Handy and others, 173–197. Rutland, Vermont: Charles E. Tuttle Co.

Pope, Willis T. 1926. Banana culture in Hawaii. *Hawaii Agricultural Experiment Station Bulletin* 55.

Pukui, Mary Kawena, and Samuel H. Elbert. 1965. *Hawaiian-English dictionary,* 3d ed. Honolulu: University Press of Hawaii.

———. 1986. *Hawaiian dictionary: Hawaiian-English, English-Hawaiian,* rev. and enlarged ed. Honolulu: University of Hawaii Press.

Pukui, Mary Kawena, E. W. Haertig, and Catherine A. Lee. 1972. *Nānā i ke kumu [Look to the source].* Honolulu: Hui hānai, Queen Lili'uokalani Children's Center.

Riley, Thomas J. 1982. Where taro is king. *Native Planters (Ho'okupu kalo)* 1 (1): 35–41.

Roberts, Helen H. 1926. *Ancient Hawaiian music.* Bernice P. Bishop Museum Bulletin 29. Honolulu. (Kraus reprint, 1971.)

Rock, J. F. 1913. *The indigenous trees of the Hawaiian Islands.* Honolulu: Privately published. (Reproduced with introduction by S. Carlquist and addendum by D. R. Herbst, 1974, Rutland, Vermont: Charles E. Tuttle Co.)

St. John, Harold. 1973. List and summary of the flowering plants in the Hawaiian Islands. *Pacific Tropical Botanical Garden Memoirs* 1. Lawai, Kauai, Hawaii.

Simmons, D. R. 1963. *Craftsmanship in Polynesia.* Otago Museum Trust Board. Dunedin, New Zealand: John McIndoe Ltd.

Stall, Edna Williamson, 1953. *The story of lauhala.* Honolulu: Paradise of the Pacific Ltd.

Stokes, John F. G. 1899. The mat sails of the Pacific. *Occasional Papers of the Bernice Pauahi Bishop Museum* 1 (2). 25–32.

———. 1906. Hawaiian nets and netting. *Bernice P. Bishop Museum Memoirs* 2 (1): 105–163.

———. 1912. The Hawaiian fish trap. *Mid-Pacific Magazine* 4:319–324.

———. 1921. Fish poisoning in the Hawaiian Islands with notes on the custom in Southern Polynesia. *Occasional Papers of the Bernice Pauahi Bishop Museum.* Vol. 7, no. 10. Honolulu.

Summers, Catherine C. 1988. *The Hawaiian grass house in*

Bishop Museum. Bishop Museum Special Publication 80. Honolulu.

———. 1990. *Hawaiian cordage.* Pacific Anthropological Records 39. Honolulu: Bishop Museum.

Thrum, Thomas. 1902. "Hawaiian calabashes." *Hawaiian Annual,* 149–154.

———. 1910. "Heiaus." *Hawaiian Annual,* 53–71.

Thurston, Lucy. 1882. *Life and times of Mrs. Lucy G. Thurston. Gathered, selected and arranged by herself.* Ann Arbor, Michigan: S. C. Andrews.

Valeri, Valerio. 1985. *Kingship and sacrifice: Ritual and society in ancient Hawaii.* Chicago: University of Chicago Press.

Vancouver, George. 1967. *Voyage of discovery to the North Pacific Ocean and around the world.* Reprint. Amsterdam: N. Israel; New York: DaCapo Press.

Wagner, Warren L., Derral R. Herbst, and S. H. Sohmer. 1990. *Manual of the flowering plants of Hawai'i.* 2 vols. Bishop Museum Special Publication 83. Honolulu: University of Hawaii Press and Bishop Museum Press.

Webb, E. Lahilahi. 1965. "Featherwork and clothing." In *Ancient Hawaiian civilization,* E. S. C. Handy and others, 135–143. Rutland, Vermont: Charles E. Tuttle Co.

Whitney, Leo D., F. A. I. Bowers, and M. Takahashi. 1939. Taro varieties. *Hawaii Agricultural Experiment Station Bulletin* 85.

Wichman, Juliet Rice. 1965. "Agriculture." In *Ancient Hawaiian civilization,"* E. S. C. Handy and others, 113–121. Rutland, Vermont: Charles E. Tuttle Co.

Winne, Jane Lathrop. 1965. "Music." In *Ancient Hawaiian civilization,* E. S. C. Handy and others, 199–211. Rutland, Vermont: Charles E. Tuttle Co.

Wise, John H. 1965a. "The history of land ownership in Hawaii." In *Ancient Hawaiian civilization,* E. S. C. Handy and others, 81–93. Rutland, Vermont: Charles E. Tuttle Co.

———. 1965b. "Food and its preparation. In *Ancient Hawaiian civilization,* E. S. C. Handy and others, 95–103. Rutland, Vermont: Charles E. Tuttle Co.

———. 1965c. "Medicine." In *Ancient Hawaiian civilization,* E. S. C. Handy and others, 257–267. Rutland, Vermont: Charles E. Tuttle Co.

Yamasake, Joanne. 1982. "Native Hawaiian dyes." Senior thesis, Bachelor's of Arts. Scripps College, Claremont, California.

INDEX

ABOUT THE AUTHOR

Beatrice H. Krauss is research affiliate at the Harold L. Lyon Arboretum and in the botany department, University of Hawaii at Mānoa. She teaches classes at the university and at the arboretum, lectures at schools and to numerous organizations, and conducts workshops for both adults and children on Hawaiian ethnobotany.

Raised in Mānoa valley, Ms. Krauss has long been recognized as an authority on native use of plants in Hawai'i. She received bachelor's and master's degrees from the University of Hawaii and, in 1988, was awarded an honorary doctorate.

Production Notes

Composition and paging were done on the Quadex Composing System and typesetting on the Compugraphic 8400 by the design and production staff of University of Hawaii Press.

The text typeface is Baskerville and the display typeface is Trump.

Offset presswork and binding were done by The Maple-Vail Book Manufacturing Group. Text paper is Glatfelter Offset Smooth, basis 60.